T0133266

And No Birds Sing

AND 🌿
NO BIRDS
SING

Rhetorical Analyses
of Rachel Carson's
Silent Spring

EDITED BY
Craig Waddell

WITH A FOREWORD BY
Paul Brooks

Southern Illinois University Press
Carbondale and Edwardsville

Printed in the United States of America
03 02 01 00   4 3 2 1
Frontispiece: Rachel Carson on the deck of her Maine cottage, summer 1962.
Photograph by Stanley Freeman. Used by permission of Stanley Freeman, Jr.

Library of Congress Cataloging-in-Publication Data

And no birds sing : rhetorical analyses of Rachel Carson's Silent spring / edited by
    Craig Waddell ; with a foreword by Paul Brooks.
        p. cm.
    Includes bibliographical references and index.
1. Carson, Rachel, 1907–1964. Silent spring. 2. Rhetoric. I. Waddell, Craig.
QH545.P4C3833 2000
363.738'4—dc21
ISBN 0-8093-2218-8 (cloth : alk. paper)   99-34724
ISBN 0-8093-2219-6 (pbk. : alk. paper)   CIP

*This book is dedicated to
Paul Brooks (1909–1998) and
Edward P. J. Corbett (1919–1998),
both of whom passed away
before the book went to print.*

The sedge is wither'd from the lake,
  And no birds sing.

—John Keats, "La Belle Dame sans Merci,"
quoted by Rachel Carson in *Silent Spring*

Contents

Foreword: Rachel Carson and *Silent Spring*

PAUL BROOKS

ﻉﻝ To write about Rachel Carson is a pleasure. Sometimes I feel that it is almost a duty: she was such a quiet, modest person that to know her from her writings alone is not easy. Some people have referred to her as shy. I should rather say that she always seemed quiet and reserved, no matter what the circumstances. (And the circumstances following the publication of her last and most influential book, *Silent Spring,* were hair-raising.) As a writer, she was very professional—a joy to work with. She did, however, have one limitation, which may have hampered her when she was being entertained as a celebrity: she had little enthusiasm for small talk. (I am reminded of a remark once made to me by a guest as we were leaving a luncheon given by Houghton Mifflin Co. in honor of Winston Churchill: "He's a very interesting man, but he hasn't much small talk, has he?")

Surely few, if any, books that have influenced our attitude toward the natural world have been undertaken so reluctantly and so courageously as *Silent Spring.* When Rachel started to work on the book in 1958, she was already a world-famous writer, thanks to a record-breaking bestseller, *The Sea Around Us,* and its companion volume, *The Edge of the Sea.* Publishers would have been eager to accept any book project that took her fancy. Rachel—as she said herself—was not, at heart, a crusader. Why then did she commit herself to such an unattractive subject as pesticides? And how was she able to make a rather specialized scientific subject a work of literature?

First, if I may, a personal note about how I came to meet Rachel when I was editor in chief of Houghton Mifflin Co. in Boston. The circumstances leading up to that meeting were somewhat bizarre. A young woman named Rosalind Wilson, daughter of the famous literary critic Edmund Wilson, had become a valuable member of Houghton Mifflin's editorial staff. During the summer months, her father liked to entertain various literary characters at his summer home on Cape Cod. One Monday morning in early summer, after attending such a party, Rosalind came into my office with an idea for a book, a book that she said was badly needed. She went on to describe the events of the weekend. There had been a storm on Saturday that had kept everyone indoors. Sunday, however, was clear and sunny: most of the guests had taken a walk along the beach. Here they had found many horseshoe crabs apparently stranded on the sand. To save the lives of these unfortunate creatures, Mr. Wilson's guests had conscientiously thrown them back into the ocean. Alas, when the guests returned for lunch, a scientifically literate gentleman (who had not accompanied them) was appalled. The crabs, he explained, were mating and laying eggs. What these presumed rescuers had done was to interfere with the next generation of horseshoe crabs! Rosalind was disturbed: this sort of ignorance was inexcusable. We should get some authority to write a book on the life of the seashore. Where could we find such a person? Neither of us knew, but we would keep the project in mind.

Soon after this conversation, I happened to be in Houghton

Mifflin's New York office when a friend and literary agent who specialized in handling "nature writers" introduced me to a quiet young woman: chief of publications for the U.S. Fish and Wildlife Service and author of a beautifully written (but commercially unsuccessful) book published in New York by Simon and Schuster, entitled *Under the Sea Wind,* with which I was well acquainted and greatly admired. Would she, I asked, be interested in writing the book on seashore life that Rosalind and I felt was so badly needed? She would. We reached a publishing agreement then and there. Such was the origin of *The Edge of the Sea,* the book that gave Rachel the most pleasure to create, involving as it did strenuous but happy hours of research along the Atlantic coast, from the rocky shores of New England to the sand beaches of the South and the mangrove forests of the subtropical zones.

Rachel was already engrossed in writing *The Edge of the Sea* when Oxford University Press published her surprise bestseller, *The Sea Around Us.* It not only topped the bestseller lists at home but was published in translation throughout the world. In due course, the seashore book, beautifully illustrated by Bob Hines (an artist for the U.S. Fish and Wildlife Service), became a companion volume to her book on the open ocean.

To most authors, this would seem like an ideal situation: an established reputation, freedom to choose one's own subject, publishers more than ready to contract for anything one wrote. It might have been assumed that her next book would be in a field that offered the same joy in research as did its predecessors. Indeed she had such projects in mind. But this was not to be.

While working for the Fish and Wildlife Service during World War II, Rachel and her scientific colleagues had become alarmed by the widespread use of DDT and other long-lasting poisons in socalled agricultural control programs. Immediately after the war, when these dangers had already been recognized, she tried in vain to interest a magazine in publishing an article on the subject. A decade later, when the spraying of pesticides and herbicides (some of them many times more toxic than DDT) was causing wholesale destruction of wildlife and its habitat, and clearly endangering human life, Rachel decided that she had to speak out. Again, she tried to interest magazines in an article. Though she was by now a well-known writer, the magazine publishers, fearing to lose advertising, turned her down. For example, a manufacturer of canned baby food claimed that such an article would cause "unwarranted fear" among mothers who used his products. (The one exception was the *New Yorker,* which would later serialize part of *Silent Spring* in advance of book publication.)

So the only answer was to write a book—book publishers being free from advertising pressure. Rachel tried to find someone else to write the book, but at last she realized that, if it were to be done, she would have to do it herself. Many of her strongest admirers questioned whether she could write a salable book on such a dreary subject. She shared their doubts, but she went ahead because she had to. "There would be no peace for me," she wrote to a friend, "if I kept silent."

The book that became *Silent Spring* was more than four years in the making. It required a very different kind of research than that for her previous books. She could no longer recount the delights of the laboratories at Woods Hole or of the marine rock pools at low tide. Joy in the subject itself had to be replaced by a sense of almost religious dedication and extraordinary courage: during the final years she was plagued with what she termed "a whole catalogue of illnesses."

Her reasons for writing the book are clear. But how did she manage to make a readable—indeed an eloquent—book on the esoteric subject of chlorinated hydrocarbons? Her early life gives a clue to the answer. Thanks to her mother, Rachel grew up with a deep appreciation of the beauty and mystery of the natural world. As she remarked many years later: "I can remember no time when I wasn't interested in the out-of-doors and the whole world of nature." And from earliest childhood, she assumed that she was going to be a writer. Beginning at age ten, she wrote stories and essays for *St. Nicholas* magazine. Her dedication to writing continued through high school and on into college, where she started out as an English major but later, thanks to a brilliant teacher, became fascinated with zoology and changed to that field. "At the time," as I wrote in my biography of her, *The House of Life: Rachel Carson at Work,*

> she believed that she had abandoned her dream of a literary career; only later did she realize that, on the contrary, she had discovered what she wanted to write about. The merging of these two powerful currents—the imagination and insight of a creative writer with a scientist's passion for fact—goes far to explain the blend of beauty and authority that was to make her books unique. (18)

With such a background, Rachel was particularly well equipped to write the book that became *Silent Spring*. By happy coincidence, at the very time we were discussing the situation, Dr. Robert Cushman Murphy, the distinguished curator of birds at the American Museum of Natural History in New York (and a friend of us both), was engaged in a lawsuit to stop the spraying of his estate on Long Island with insecticide. The lawsuit eventually failed. But the technical information on which it was based, gathered from experts all over the world, became the original core of research for *Silent Spring*. And when *Silent Spring* was published four years later, Dr. Murphy became one of its strongest supporters.

It was a letter to Bob Murphy that suggested to me the title for the book that Rachel was writing—and sending to me, chapter by chapter. A woman in the Midwest whose land had been sprayed, without her permission, with one of these powerful insecticides complained to Bob that all bird song had vanished: spring now came silently. I originally suggested "Silent Spring" as the title for Rachel's chapter about the effects of these poisons on birds. But when the manuscript was at last completed, and we still had no title, it occurred to me that, metaphorically, *Silent Spring* applied to the book as a whole. Not all of my colleagues at Houghton Mifflin agreed. "Blind title," complained one of them, "Doesn't mean anything." Fortunately, it did mean something to Rachel and has to readers throughout the world.

When *Silent Spring* was published in late September 1962, Rachel was well aware of the storm that it would arouse in the pesticide industry. The attacks began even before the book's publication, when extracts appeared in the *New Yorker.* The National Agricultural Chemical Association spent a quarter of a million dollars to discredit the book in the press and on television. The result was more publicity than Houghton Mifflin could possibly have afforded. One of the major chemical companies made a unique contribution. Ignoring facts altogether, it ridiculed the book in a parody entitled *The Desolate Year,* depicting the horrors of a world without pesticides, something that Rachel Carson had specifically *not* recommended. A trade magazine depicted a future in which people would be reduced to feeding on acorns. But the showcase of the campaign was the television commercial featuring a sinister figure in a white lab coat, gloomily recounting the lethal effects of doing without these poisons. He was a good actor; he provided his audience with a chilling experience. Yet I doubt that many viewers accepted his claim that *Silent Spring* was leading us down the path to destruction.

Rachel Carson was modest about her accomplishment. As she wrote to a close friend when the manuscript of *Silent Spring* was nearing completion: "The beauty of the living world I was trying to save has always been uppermost in my mind—that, and anger at the senseless, brutish things that were being done. . . . Now I can believe I have at least helped a little." In fact, her book helped to make *ecology*, which was an unfamiliar word in those days, one of the greatest popular causes of our time. It led to environmental legislation at every level of government.

As I wrote of Rachel in my biography of her,

> Though she had the broad view of the ecologist who studies the infinitely complex web of relationships between living things and their environment, she did not concern herself exclusively with the great impersonal forces of nature. She felt a spiritual as well as a physical closeness to the individual creatures about whom she wrote: a sense of identification that is an essential element in her literary style. (8)

One thinks of Henry Thoreau, who felt wiser in all respects for knowing that there was a minnow in the brook: "Methinks I have need even of his sympathy, and to be his fellow in a degree." If I had to choose a single revealing moment during my long friendship with Rachel, it would be shortly after dusk one July evening at her Maine cottage, while she was working on *The Edge of the Sea*. We had spent an hour after supper examining minute sea creatures under her brightly lit binocular microscope: tube worms, rhythmically projecting and withdrawing their pink, fanlike tentacles in search of invisible food; tiny snails on fronds of seaweed; flowerlike hydroids; green sponges, whose ancestry goes back to the earliest record of life on earth. At last we were finished. Then, pail and flashlight in hand, she stepped carefully over the kelp-covered rocks to return the living creatures to their home. This, I think, is what Albert Schweitzer (to whom *Silent Spring* is dedicated) meant by "reverence for life." In one form or another, it lies behind everything that Rachel Carson wrote.

Thirty-six years after its original publication, *Silent Spring* has more than a historical interest. Rachel Carson was a realistic, well-trained scientist who possessed the insight and sensitivity of a poet. She had an emotional response to nature for which she did not apologize. The more she learned, the greater grew what she termed "the sense of wonder." So, with *Silent Spring,* she succeeded in making a book about death a celebration of life. To appreciate her achievement, one should bear in mind that, in her intense feeling for our relationship to the living world around us, she was ahead of her time. When she began writing, the term *environment* had few of the connotations it has today. Conservation was not yet a political force. To the public at large, the word *ecology*—derived from the Greek for *habitation*—was unknown, as was the concept for which it stood.

Rereading *Silent Spring* today, one is aware that its implications are far broader than the immediate crisis with which it dealt. By awaking us to a specific danger—the poisoning of the earth with chemicals—Rachel Carson has helped us to recognize many other ways (some little known in her time) in which humankind is degrading the quality of life on our planet. And *Silent Spring* will continue to remind us that in our overorganized and overmechanized age, individual initiative and courage still count: change can be brought about not through incitement to war or violent revolution but rather by altering the direction of our thinking about the world in which we live.

Acknowledgments

𝄢 The editor and contributors gratefully acknowledge use of the Rachel Carson Papers, The Yale Collection of American Literature, Beinecke Rare Book and Manuscript Library, Yale University.

Excerpts from *Silent Spring* by Rachel Carson. Copyright © 1962 by Rachel L. Carson, renewed 1990 by Roger Christie. Reprinted by permission of Houghton Mifflin Co. All rights reserved.

Excerpts from unpublished and oral Rachel Carson material. Copyright © 1998 by Roger Christie. Reprinted by permission of Frances Collin, Trustee u-w-o Rachel Carson. All rights reserved.

Excerpts from *The House of Life* by Paul Brooks. Copyright © 1972 by Paul Brooks. Reprinted by permission of Frances Collin, Literary Agent.

Excerpts from *Always, Rachel* edited by Martha Freeman. Courtesy Beacon Press.

And No Birds Sing

1

CRAIG WADDELL

The Reception of *Silent Spring:* An Introduction

❧ Ernest Hemingway once wrote that "[a]ll modern American literature comes from one book by Mark Twain called *Huckleberry Finn*." It would not be too much of an exaggeration to make a similar claim for *Silent Spring*'s relationship to the modern American environmental movement. Although the American environmental movement traces its roots to such nineteenth-century visionaries as Henry David Thoreau, George Perkins Marsh, and John Muir—all of whom were concerned with the preservation of wilderness—the modern environmental movement—with its emphasis on pollution and the general degradation of the quality of life on the planet—may fairly be said to have begun with one book by Rachel Carson called *Silent Spring*.

Environmental historians are nearly unanimous on this point. Consider, for example, the following accolades (see also Briggs 62; Ehrlich, "Paul Ehrlich" 34; Hays 52; Hynes 9; Killingsworth and Palmer, *Ecospeak* 65; Lear, "Rachel Carson's *Silent Spring*" 23; Merideth 21–22; and Oravec 77):

One hundred and ten years after *Uncle Tom's Cabin* Rachel Carson wrote another book that exploded against traditional American assumptions. . . . *Silent Spring* was a landmark in the development of an ecological perspective. It did much to accelerate the new environmentalism and generated the most widespread public consideration of environmental ethics to that date. (Nash 78)

Silent Spring became one of the seminal volumes in conservation history: the *Uncle Tom's Cabin* of modern environmentalism. (Fox 292)

Silent Spring by Rachel Carson . . . is now recognized as one of the truly important books of this century. More than any other, it changed the way Americans, and people around the world, looked at the reckless way we live on this planet. (Shabecoff 107)

Since it was first published in 1962, Carson's eloquent and well-documented exposé on the indiscriminate use of pesticides has never been out of print. Her work led to a special investigation by President John F. Kennedy's Science Advisory Committee, which confirmed her conclusions, and eventually led to the formation of the Environmental Protection Agency (EPA) and the banning of DDT. In his introduction to the 1994 edition of *Silent Spring*, Vice President Al Gore notes that "in 1992, a panel of distinguished Americans selected *Silent Spring* as the most influential book of the last fifty years" (xviii). Continued interest in *Silent Spring* is also suggested by the growth of the Rachel Carson Council, whose newsletter is distributed to more than seven thousand members; by the 1993 PBS documentary "Rachel Carson's *Silent Spring*"; by Martha Freeman's 1995 publication of *Always, Rachel: The Letters of Rachel Carson and Dorothy Freeman;* by the 1996 publication of Branda

Miller's CD-ROM *Witness to the Future: A Call for Environmental Action,* which includes the first electronic version of *Silent Spring;* and by Linda Lear's 1997 biography *Rachel Carson: Witness for Nature.*

Although Aristotle argued that rhetoric concerns itself with "the things about which we make decisions, and into which therefore we inquire" (1357a.25)—and although *Silent Spring* provides a striking example of such discourse—rhetoricians have thus far made few attempts to analyze or explain *Silent Spring's* phenomenal impact on public deliberation about the environment. This volume is devoted to such analyses.

Rachel Carson's Life and Work

Rachel Carson was born on 27 May 1907, in Springdale, Pennsylvania, a landlocked community eighteen miles northeast of Pittsburgh, far from the Atlantic Ocean that would eventually have such a profound impact on her life.[1] Her father dreamed of enhancing the family's financial fortunes by developing a sixty-four-acre tract of woods and fields he had purchased on the outskirts of town (Brooks 15–16; Lear, *Rachel Carson* 10). Although her father never realized his vision, this land did provide Rachel with an early acquaintance with the natural world. A precocious child who was often kept home from school by her mother, Rachel explored this land with her mother and with her older brother and sister. From an early age, she developed a love of books—a love cultivated by her mother, a graduate of the Washington Female Seminary in Washington, Pennsylvania—and she became especially fascinated with books about the sea. Despite her frequent absences from class, between her mother's tutoring and her own penchant for reading, Rachel never fell behind in her schoolwork (Brooks 16; Lear, *Rachel Carson* 21).

Even as a child, Rachel aspired to be a writer, and in 1918, at age eleven, she published her first piece, an essay in *St. Nicholas,* a children's magazine. After graduating from high school, she entered the Pennsylvania College for Women (later renamed Chatham College), initially majoring in English. In her junior year, however—under the influence of Mary Scott Skinker, her charismatic biology professor—Carson changed her major to zoology. With this combination of majors, she was able to blend the two loves of her life: writing and nature (Brooks 17–18). With Skinker's encouragement, Carson applied and was accepted into graduate school at Johns Hopkins University. Before beginning at Johns Hopkins in the fall of 1929, she completed a summer fellowship at the Marine Biological Laboratory at Woods Hole, Massachusetts, where she saw the ocean for the first time (see McCay 9). This experience was to have a profound impact on her career (Lear, *Rachel Carson* 60).

Carson received her master's degree in marine zoology from Johns Hopkins in 1932. For the next several years, she taught at Johns Hopkins and at the University of Maryland. However, in 1935, when her father died suddenly in the midst of the Depression, increased family responsibilities forced her to seek more profitable employment. She applied for and received a temporary position as a writer with the U.S. Bureau of Fisheries (renamed in 1940 the U.S. Fish and Wildlife Service), and in 1936, she obtained a full-time position with the bureau (Brooks 19–20). One of her first assignments was to write a piece about the sea for one of the bureau's radio broadcasts. Although her supervisor rejected the resulting essay, he encouraged her to submit the piece to the *Atlantic Monthly*. Carson followed this advice, and in September 1937, the *Atlantic* published her essay "Undersea." Shortly thereafter, Quincy Howe, the senior editor of Simon and Schuster, wrote to Carson asking if she planned a book on this subject (Lear, *Rachel Carson* 88–90). Although she had not previously considered a book, with Howe's encouragement, Carson pursued this idea. The result was *Under the Sea Wind*, a compelling narrative description of the life of the shore, the open sea, and the sea bottom (Brooks 34). The book was published on 1 November 1941, six weeks prior to the bombing of Pearl Harbor; and although it received a good initial response, the country's preoccupation with the war soon eclipsed its publication (Lear, *Rachel Carson* 105; McCay 33–34). Carson continued with the Fish and Wildlife Service during the war; and after learning through her work of the potentially devastating effects of widespread use of DDT, in 1945 she proposed to *Reader's Digest* an article on that subject. Failing to interest the *Digest* in such an article, she turned to other work, and in 1949, she was appointed chief editor of all Fish and Wildlife Service publications (Lear, *Rachel Carson* 148). However, as Lear notes, "At no time after [1945] was [Carson] ever dispassionate about the issue of synthetic chemical pesticides" (*Witness* 312).

In 1951, the *New Yorker* serialized and Oxford University Press published the book that made Rachel Carson famous: *The Sea Around Us*. This book conveyed gracefully and for nontechnical readers much of what had been learned about the oceans during the war years. It emphasized our dependence on the oceans and Carson's belief that we would become even more dependent as we destroyed the land (Brooks 110; Carson, Preface 13–18). The book stayed on the *New York Times* bestseller list for eighty-six weeks, setting a new record (Brooks 127, 129–30). The phenomenal success of *The Sea Around Us* led to the republication (by Oxford) of Carson's first book, which quickly joined her second book on the bestseller list. In 1952, the success of these publications allowed Carson to resign from the Fish and Wildlife Service in order to devote her full time to writing. It also allowed her to purchase a home in Maryland and a summer cottage in Southport, along her beloved Maine coast (Brooks 159; Lear, *Rachel Carson* 324–35).

Carson's next project, already underway when *The Sea Around Us* was published, was a literary take on a field guide to the Atlantic coast of the United States. *The Edge of the Sea,* published in 1955 by Houghton Mifflin, also became a bestseller (McCay 59). Early in

1957, Carson's niece Marjorie died, leaving a five-year-old son, whom Carson adopted (Freeman 216, n. 3). Carson herself suffered from a host of ailments that continually competed for her attention and energies. Nevertheless, despite her family responsibilities and ill health, she continued with her work. By the fall of 1957, she had a series of publications in mind when two cases drew her attention once more to the potentially devastating effects of indiscriminate use of synthetic pesticides: the U.S. Department of Agriculture's program to eradicate the fire ant and a Long Island suit seeking to stop aerial spraying of private land with DDT (Lear, *Rachel Carson* 312). In January 1958, Carson's friend Olga Huckins wrote a letter to the editor of the *Boston Herald* and sent a copy of the letter to Carson. Huckins described the horrible deaths of birds in her private bird sanctuary after—against her wishes—the Commonwealth of Massachusetts saturated her property with DDT in an aerial spraying campaign designed to control mosquitoes. She asked Carson about people in Washington who might be able to help. Failing to find anyone else who would and could write an exposé on this subject, Carson decided to undertake the task herself. Her first intention was to produce a magazine article; but finding magazines generally unresponsive and, in any case, the subject too expansive for such an article, she proposed publishing a book on indiscriminate use of pesticides with Houghton Mifflin and serializing significant chapters in the *New Yorker.* During her early research, she received invaluable assistance from the vast body of expert testimony that had accumulated during a 1957–58 lawsuit brought by Robert Cushman Murphy—renowned curator of birds at the American Museum of Natural History—and others. This suit sought to prevent the state and federal governments from aerial spraying on Long Island in their campaign to eradicate the gypsy moth. The case went all the way to the Supreme Court, which refused to hear it on the basis of a technicality. Although the suit failed, it provided Carson with a wealth of relevant scientific material. Her literary reputation and her many years' experience as an editor for the Fish and Wildlife Service provided her with the additional access she needed to research this complex and controversial subject (see Brooks 227–43; Lear, *Rachel Carson* 313).

For the next four and one-half years, Carson persisted in her task, despite personal tragedies—including the 1958 death of her mother—and an oppressive battery of health problems. Finally, in June 1962, the *New Yorker* serialized ten chapters of *Silent Spring* over three successive issues; and on 27 September 1962, Houghton Mifflin published the full text. The response to both publications was overwhelming. As Brooks says, "Perhaps not since the classic controversy over Charles Darwin's *The Origin of Species* . . . had a single book been more bitterly attacked by those who felt their interests threatened" (293). The National Agricultural Chemical Association led the attack, committing a quarter of a million dollars to improving the image of the industry and refuting Carson's case against the indiscriminate use of chemical insecticides (Brooks 294); the Manufacturing Chemists' Association also attacked the book (Lear, *Rachel Carson* 413). The response from the reading public, on the other hand, was overwhelmingly positive. By December, more than one hundred thousand copies had been sold in bookstores. In addition, the Book-of-the-Month Club chose *Silent Spring* as its main selection for October, and the Consumer's Union ordered a special paperbound edition for its members (Graham 69).[2] By the following spring, sales had passed the half-million mark (Gore xvii). On 3 April 1963, *CBS Reports* aired "The Silent Spring of Rachel Carson," which led to a dramatic escalation of the debate (Lear, "Rachel Carson's *Silent Spring*" 39). Shortly after the *New Yorker* serialized key chapters of *Silent Spring,* President Kennedy mentioned in a press conference that his administration was examining the concerns Carson raised. On 15 May 1963, the President's Science Advisory Committee issued its report, *Use of Pesticides.* According to the editors of *Science,* this report amounted to "a fairly thorough-going vindication" of *Silent Spring* (qtd. in Graham 79). The report concluded that "[p]ublic literature and the experiences of Panel members indicate that, until the publication of 'Silent Spring' by Rachel Carson, people were generally unaware of the toxicity of pesticides" (23).

Rachel Carson died of cancer and heart disease on 14 April 1964, less than two years after the publication of *Silent Spring* (Brooks 327; Lear, *Rachel Carson* 480). As Eric Sevareid noted shortly after the publication of the President's Science Advisory Committee's report, Carson had achieved her two immediate aims: alerting the public and building a fire under the government (see Graham 79). The formation of the EPA in 1970 and the EPA's 1972 ban on the use of DDT (with the significant exceptions of quarantine and public-health uses and manufacture for export) can be traced directly to Carson's work (Dunlap 234–36). In 1980, President Jimmy Carter posthumously awarded Rachel Carson the Presidential Medal of Freedom, the highest honor the country can bestow on a civilian (Gartner 28). The citation for this award reads in part, "Always concerned, always eloquent, she created a tide of environmental consciousness that has not ebbed" (qtd. in Gartner 28). Despite her accomplishments, of course, Rachel Carson could not single-handedly ensure sound environmental policy. As Vice President Gore points out, "Since the publication of *Silent Spring,* pesticide use on farms alone has doubled to 1.1 billion tons a year, and production of these dangerous chemicals has increased by 400 percent" (xix). Despite this increased use of pesticides, crops lost to pests in the United States increased from 32 percent in 1945 to 37 percent in 1984 (Orr 54). Clearly, the task Carson began remains unfinished. As she said herself after completing her manuscript, "I *had* done what I could—I had been able to complete it—now it had its own life" (Freeman 394).

Explaining Public Reception of *Silent Spring*

Various efforts to explain the immense impact of *Silent Spring* have focused on what I will call hypothesis A, or the apocalyptic thesis; that is, that the success of *Silent Spring* is attributable to the appeal of Carson's apocalyptic vision, which is especially prominent in "A Fable for Tomorrow," the book's opening chapter. For example, Killingsworth and Palmer suggest that "Carson's chief rhetorical strategy was the development of startling contrasts and dramatically rendered conflicts in a future-oriented report on the environment, a kind of apocalyptic narrative befitting a latter-day prophet" (*Ecospeak* 65; see also Bercovitch for an extended analysis of the role of apocalyptic vision in American history).

I began my own study of public response to *Silent Spring* with what I will call hypothesis B, or the zeitgeist thesis: that is, that it was not Carson's apocalyptic appeal per se that facilitated public reception of her work but rather her ability to ally her concerns with the spirit of the times—which happened to be dominated by an apocalyptic fear of nuclear holocaust.[3] This hypothesis suggests that contemporary advocates who would learn from Carson's example should focus not on the construction of apocalyptic narratives per se—as many environmentalists, such as Paul Ehrlich (*Population Bomb*) and Stephen Schneider (*Global Warming*), have done— but on allying their appeals with contemporary concerns (compare Killingsworth and Palmer, "Discourse" 11). This point was not lost on Carson's opponents, some of whom allied their attacks on Carson with prevailing concerns about world Communism (see Graham 49). Hence, in the post–Cold War era, rather than generating apocalyptic visions, environmental advocates might more profitably ally their appeals with prevailing concerns, such as human health and economic issues (which may or may not be apocalyptic).

Hypothesis B guided my preliminary research with the Rachel Carson Papers at the Beinecke Rare Book and Manuscript Library at Yale. In working with these papers, I gave particular attention to editorial correspondence, promotional materials, and reviews; I also examined letters to editors, responses from chemical companies, and government documents. I borrowed my preliminary research assumption from I. A. Richards, who, in *Practical Criticism,* suggests that, in responding to poetry, his students typically project into the poetry—and hence, include in their responses to it—significant features of the culture in which they are immersed. He goes on to suggest, therefore, that *Practical Criticism* is, in part, "a record of a piece of fieldwork in comparative ideology" (6). That is, he suggests that he has used responses to poetry as a projective test to reveal contemporary beliefs and values within his cultural context. Book reviews can provide similar insights: in responding to a text such as *Silent Spring,* reviewers inevitably reveal some of their culture's prevailing concerns and biases, and in doing so, they reveal with which aspects of the spirit of the times the text under review has reverberated. Hence, if hypothesis B were valid, one might expect to find in reviews of *Silent Spring* repeated allusions to the Cold War, nuclear weapons, radiation, and nuclear holocaust.

In reading through the approximately six hundred reviews of *Silent Spring* included in the Carson Papers, I did find such references.[4] However, they cannot fairly be said to be a dominant feature of the reviews. Instead, the reviews include a diverse range of recurring references, including the following: (1) comments on Carson's impressive data (e.g., her fifty-five-page list of sources) and her scientific credentials; (2) comments on Carson's literary grace; (3) comments on Carson's moderation and realism; (4) comments on the service Carson provided by informing the public; (5) comments on Carson's concern for human health; (6) references to serialization in the *New Yorker;* (7) comments on related fears, such as radiation, the cranberry scare of 1959, and thalidomide (Graham notes that "[p]art of *Silent Spring's* impact on the public that summer resulted from the coincidence that the *New Yorker* articles followed almost immediately the thalidomide tragedy" [50]; see also

Lear, *Rachel Carson* 411); (8) claims that chemical sprays are not working (e.g., not saving the elms); and (9) superlatives (e.g., "It is the most important book that I have read for review"). These nine patterns interact and combine to privilege some readings of *Silent Spring* more than others—that is, they define certain cultural experiences or expectations that forestructure responses to the text.[5]

This range of recurring references suggests a third hypothesis, hypothesis C, or the overdetermination thesis: that is, that no one factor—either within or outside the text—can adequately explain the success of *Silent Spring*. A variety of natural "controls" support this hypothesis. For example, the modest attention paid to three books addressing the same subject—Murray Bookchin's *Our Synthetic Environment,* published six months prior to *Silent Spring* (see Hynes 3); Theron G. Randolph's *Human Ecology and Susceptibility to the Chemical Environment,* also published in 1962; and Robert Rudd's *Pesticides and the Living Landscape,* published one and one-half years after *Silent Spring*—challenges the hypothesis that the spirit of the times alone explains *Silent Spring*'s success.[6] Likewise, despite Bookchin's claim that the world was "captivated by the superb prose of *Silent Spring*" (qtd. in Hynes 3), the lackluster sales of Carson's first book, *Under the Sea Wind,* published just six weeks prior to the bombing of Pearl Harbor, challenge the hypothesis that Carson's literary grace was a sufficient condition for *Silent Spring*'s success. (Despite the poor sales, Carson believed *Under the Sea Wind* to be one of her most creative works [see Brooks 35].)

Mary Hesse argues that "[t]heories are logically constrained by facts, but are underdetermined by them: that is, while, to be acceptable, theories should be more or less plausibly coherent with facts, they can be neither conclusively refuted nor uniquely derived from statements of fact alone" (187). Likewise, public reception of *Silent Spring* was underdetermined both by Carson's apocalyptic vision and by the apocalyptic spirit of the times with which this vision was allied; that is, public reception was partly determined by these factors and partly determined by other factors within the text, by other texts, and by other aspects of the historical context within which *Silent Spring* was read. Hence, analysis of public reception of *Silent Spring* must consider not only multiple factors within the text but also the influence of other texts and of the historical context. Framed in terms of the concept of *overdetermination*—which Raymond Williams defines as "determination by multiple factors" (83)—to claim that the reception of a text is underdetermined by X is to claim that it is overdetermined in that it is partly determined by X and partly determined by other factors. According to Williams, "[T]he concept of *overdetermination* is more useful than any other as a way of understanding historically lived situations and the authentic complexities of practice" (88).

Hypothesis C, then, suggests a different approach to understanding and learning from the public response to *Silent Spring*. Rather than looking for the single "silver bullet" that might first explain public reception of *Silent Spring* and second be used to model further environmental discourse, we should look to diverse contributing factors that collectively overdetermine such a response. Prominent among these might still be allying our concerns with the spirit of the times, but this alone cannot explain Carson's success, and it should not serve as a singular model for those who would follow in her path. The diverse approaches to *Silent Spring* taken by the contributors to the current volume only begin to suggest the diverse lessons that might be drawn from Carson's phenomenally successful book.

Notes

1. In this section, I draw primarily on *The House of Life: Rachel Carson at Work*, by Paul Brooks, and on *Rachel Carson: Witness for Nature*, by Linda Lear, but also on the historical and biographical writings of Dunlap, Gartner, Gore, Graham, Hynes, McCay, Oravec, and others, and on my own research with the

Rachel Carson Papers in Yale University's Beinecke Rare Book and Manuscript Library.

2. Writing in the September 1962 edition of *Book-of-the-Month Club News*, Club Chairman Harry Scherman said,

> Only with two Selections in recent years have we advised members that it was unwise to use their privilege of rejection or substitution. We do so once again in the case of this remarkably illuminating new book by Rachel Carson, for it is certain to be history-making in its influence upon thought and public policy over the world.

3. Carson not only appealed to the widespread public *fear* of nuclear war and nuclear radiation, she also capitalized on widespread public *understanding* of such things. In doing so, she was able to take advantage of the government's expansive civil-defense information campaigns and base her explanation of how pesticides could impact human health on what the public already knew about radiation. For example, she says that "chemicals are the sinister and little-recognized partners of radiation in changing the very nature of the world" (6). She goes on to explain how strontium 90 can be spread throughout the environment and accumulate in human bodies—information that should already be familiar to her readers from the civil-defense campaigns—and then makes a point-by-point comparison with chemical pesticides. (See Ralph H. Lutts's essay, this volume.)

4. Most of these reviews are from 1962–63, and most were collected for Carson and her agent, Marie Rodell, by the Romeike Press Clippings Service.

5. I found some of these same patterns in the editorial correspondence and promotional materials and in letters to editors. Such correlations suggest—but do not establish—cause-and-effect relationships; that is, they suggest that Carson, her agent, and her publisher explicitly employed such devices to structure the response to *Silent Spring*. Further evidence in the correspondence suggests that, for at least some of these factors, this was the case. Hence, beyond the influence of the book itself, public reception of *Silent Spring* was shaped (1) by Carson's correspondence with agents, editors, and publicists; (2) by publicists' communication with reviewers, columnists, and editors; and (3) by reviewers', columnists', and editors' communication with the general public. In this process, Carson underdetermined (i.e., structured, shaped, influenced, or constrained, but did not determine) the content and nature of the promotional materials; the promotional materials underdetermined the nature

of the reviews, columns, and editorials; and the reviews, columns, and editorials underdetermined public reception of *Silent Spring*. All three of these features were also underdetermined by the historical context within which they were immersed, suggesting, again, the value of allying appeals with prevailing concerns. (Promotional materials underdetermined by Carson also include those within the book, such as the cover, title, illustrations, and list of principal sources.)

6. Linda Lear notes, however, that William F. Longgood's 1960 exposé, *The Poisons in Your Food*, did achieve some popular success (*Witness* 371).

Works Cited

Aristotle. *Rhetoric*. Trans. W. Rhys Roberts. New York: Modern, 1954.

Bercovitch, Scavan. *The American Jeremiad*. Madison: U of Wisconsin P, 1978.

Bookchin, Murray [Lewis Herber] *Our Synthetic Environment*. New York: Knopf, 1962.

Briggs, Shirley A. "Remembering Rachel." *EPA Journal* 18.2 (May–June 1992): 62.

Brooks, Paul. *The House of Life: Rachel Carson at Work*. Boston: Houghton, 1972.

Carson, Rachel. Papers. Yale Collection of American Literature. Beinecke Rare Book and Manuscript Library, Yale University.

———. Preface. *The Sea Around Us*. Rev. ed. New York: Oxford UP, 1961.

———. *Silent Spring*. Boston: Houghton, 1962.

Dunlap, Thomas R. *DDT: Scientists, Citizens, and Public Policy*. Princeton: Princeton UP, 1981.

Ehrlich, Paul R. "Paul Ehrlich Reconsiders *Silent Spring*." *Bulletin of the Atomic Scientists* 35 (Oct. 1979): 34–36.

———. *The Population Bomb*. New York: Ballantine, 1968.

Fox, Stephen. *The American Conservation Movement: John Muir and His Legacy*. Madison: U of Wisconsin P, 1981.

Freeman, Martha, ed. *Always, Rachel: The Letters of Rachel Carson and Dorothy Freeman, 1952–1964*. Boston: Beacon, 1995.

Gartner, Carol B. *Rachel Carson*. New York: Ungar, 1983.

Gore, Al. Introduction. *Silent Spring*. By Rachel Carson. Boston: Houghton, 1994. xv–xxvi.

Graham, Frank, Jr. *Since Silent Spring*. Boston: Houghton, 1970.

Hays, Samuel P. *Beauty, Health, and Permanence: Environmental Politics in the United States, 1955–1985*. Cambridge, UK: Cambridge UP, 1987.

Hemingway, Ernest. *The Green Hills of Africa*. New York: Macmillan, 1935.

Hesse, Mary. *Revolutions and Reconstructions in the Philosophy of Science*. Bloomington: Indiana UP, 1980.

Hynes, H. Patricia. *The Recurring Silent Spring*. New York: Pergamon, 1989.

Killingsworth, M. Jimmie, and Jacqueline S. Palmer. "The Discourse of 'Environmentalist Hysteria.'" *Quarterly Journal of Speech* 81 (1995): 1–19.

———. *Ecospeak: Rhetoric and Environmental Politics in America*. Carbondale: Southern Illinois UP, 1992.

Lear, Linda J. "*Rachel Carson's Silent Spring*." *Environmental History Review* 17.2 (1993): 23–48.

———. *Rachel Carson: Witness for Nature*. New York: Holt, 1997.

Longgood, William F. *The Poisons in Your Food*. New York: Simon, 1960.

McCay, Mary A. *Rachel Carson*. New York: Twayne, 1983.

Merideth, Robert. *The Environmentalist's Bookshelf: A Guide to the Best Books*. New York: Hall, 1993.

Miller, Branda. *Witness to the Future: A Call for Environmental Action*. CD-ROM. Irvington, NY: Voyager, 1996.

Nash, Roderick Frazier. *The Rights of Nature: A History of Environmental Ethics*. Madison: U of Wisconsin P, 1989.

Oravec, Christine. "Rachel Louise Carson (1907–1964): Author, Naturalist, Environmental Advocate." *Women Public Speakers in the United States, 1925–1993: A Bio-critical Sourcebook*. Ed. Karlyn Kohrs Campbell. Westport, CT: Greenwood, 1994. 72–89.

Orr, David W. *Ecological Literacy: Education and the Transition to a Postmodern World*. Albany: State U of New York P, 1992.

President's Science Advisory Committee. *Use of Pesticides*. Washington: GPO, 1963.

"Rachel Carson's *Silent Spring*." *The American Experience*. Writ. and prod. Neil Goodwin. PBS. Peace River Films, Cambridge. 8 Feb. 1993.

Randolph, Theron G. *Human Ecology and Susceptibility to the Chemical Environment*. Springfield, IL: Thomas, 1962.

Richards, I. A. *Practical Criticism: A Study of Literary Judgment*. New York: Harcourt, 1929.

Rudd, Robert L. *Pesticides and the Living Landscape*. Madison, U of Wisconsin P, 1964.

Scherman, Harry. *Book-of-the-Month Club News*. Sept. 1962.

Schneider, Stephen. *Global Warming: Are We Entering the Greenhouse Century?* San Francisco: Sierra Club, 1989.

Shabecoff, Philip. *A Fierce Green Fire: The American Environmental Movement.* New York: Hill, 1993.

Williams, Raymond. *Marxism and Literature.* New York: Oxford UP, 1977.

2 RALPH H. LUTTS

Chemical Fallout: *Silent Spring,* Radioactive Fallout, and the Environmental Movement

❧ The landmark book *Silent Spring* played a vitally important role in stimulating the contemporary environmental movement. Never before or since has a book been so successful in alerting the public to a major environmental pollutant, rooting the alert in a deeply ecological perception of the issues, and promoting major public, private, and governmental initiatives to correct the problem. It was exceptional in its ability to combine a grim warning about pesticide poisoning with a text that celebrated the living world. *Silent Spring* has been compared in its social impact to *Uncle Tom's Cabin* (United States, *Interagency Coordination* 220–21); John Kenneth Galbraith described it as one of the most important books of Western literature ("Immortal Nominations" 13); and Robert Downs listed it as one of the "books that changed America" (260–61).[1]

This is a revised version of an essay that first appeared in *Environmental Review* 9 (1985): 210–25. Copyright © 1985, 1998 Ralph H. Lutts. Used by permission of the author.

Rachel Carson's case against the indiscriminate use of pesticides prevailed in the face of powerful, well-financed opposition by the agricultural and chemical industries. Despite this opposition, she prompted national action to regulate pesticides by mobilizing a concerned public. The book established a broad constituency for addressing the problem—broader, perhaps, than that enjoyed by any previous environmental issue. Never before had so diverse a body of people—from bird-watchers, to wildlife managers and public-health professionals, to suburban home owners—been joined together to deal with a common national and international environmental threat. Her success in the face of what might have been overwhelming opposition suggests there was something significantly different between the response to *Silent Spring* in 1962 and the pesticide-control efforts of the first half of the century.

The issue of pesticide pollution was not new. Since the introduction of Paris green around 1867, highly toxic compounds of lead and arsenic were widely used in agriculture despite the significant health hazards they presented. As one example, seventy-five million pounds of lead arsenate were applied within the United States in 1944; eight million pounds were even used in the 1961–62 crop year when DDT was preeminent. In the early decades of their use, these toxic chemicals could sometimes be found as visible coatings on farm produce in retail markets. Over the years, stories of acute poisonings and warnings of the dangers of chronic toxicity appeared in the press. Everyone was warned to scrub or peel fruits and vegetables before they were eaten. Many public-health officials attempted to institute strong regulations and strict residue tolerances, but the general public, medical profession, and agriculture industry showed only limited concern (Whorton 178; Whitaker 378; United States, *Environmental Hazards* 13). This relative indifference to the hazards of pesticides in the first half of the century stands in stark contrast to the vocal outcry following the publication of *Silent Spring*.[2]

Why is it that the book's publication in 1962 had such a major impact upon the public? The answer to this question might reveal a great deal about the origins of contemporary environmental concerns, but no one has examined it systematically. A number of answers have been suggested, focusing most often upon Carson's extraordinary skill and reputation as a writer, the general circumstances surrounding the rise of pesticide use and misuse, the publisher's marketing strategy, and the chemical industry's response. Many authors have also noted the growing public awareness of a variety of environmental problems, including water and air pollution. One of the major events to bring the hazards of pesticides to public attention was the "cranberry scare" of 1959 when people were warned against eating this traditional fruit during the Thanksgiving season because of pesticide contamination. The thalidomide syndrome also came to the public's attention shortly before the publication of *Silent Spring,* and the pictures of the distorted infant limbs caused by a supposedly beneficial drug certainly made people pay greater attention to Carson's message (Brooks, *House* 261; Graham 50–51; Taussig).

There was another issue, however, that played an equal or greater role in preparing the public to accept Carson's warning—an issue that has been largely overlooked.[3] She was sounding an alarm about a kind of pollution that was invisible to the senses; could be transported great distances, perhaps globally; could accumulate over time in body tissues; could produce chronic, as well as acute, poisoning; and could result in cancer, birth defects, and genetic mutations that may not become evident until years or decades after exposure. Government officials, she also argued, were not taking the steps necessary to control this pollution and protect the public. Chemical pesticides were not the only form of pollution fitting this description. Another form, far better known to the public at the time, was radioactive fallout. Pesticides could be understood as another form of fallout.

People in the United States and throughout the world were prepared, or preeducated, to understand the basic concepts underlying Rachel Carson's *Silent Spring* by the decade-long debate over radioactive fallout preceding it. They had already learned that poisons, in this case radioactive ones, could create a lasting global danger. To understand the deep impact of this debate upon the public, we must review the history of the fallout controversy.

The Beginning

During the heady days of the late 1940s, when the United States was the only nation possessing the atomic bomb, Americans did not worry much about this symbol of international status and power. The major cultural contribution of the 1946 U.S. A-bomb tests at the Bikini Atoll in the Pacific was the name of a new French bathing suit (Eherhart). The Soviet Union's detonation of its own atomic bomb in 1949 destroyed this complacency, and the postwar nuclear arms race began. In 1951, the U.S.S.R. exploded another two devices and the United States sixteen. By the end of 1953, both nations had conducted a total of twenty-six more tests. In November 1952, the U.S. government exploded the world's first thermonuclear device, followed by the Soviets' detonation of their own device in August 1953. In March 1954, the United States tested its first portable superbomb (*SIPRI: 1968/69* 242; Fowler 16, 209).

By the early 1950s, the public was extraordinarily interested in atomic weapons. This early interest reflected nationalistic pride, fear of the Soviets, and fascination with the bombs and the mysteries of radioactivity rather than a major concern about public health. The majority of United States A-bomb tests were conducted in Nevada, and the resulting clouds of radioactive materials, which passed over populated areas of the nation, led to growing public anxiety despite reassuring statements by the Atomic Energy Commission (AEC; see "AEC Fifth Semiannual Report"). In March 1953, for example, a *New York Times* writer reported that the AEC had determined there was no danger to American cities from the tests. The explanation may not have instilled great confidence, however. "Radioactivity in the atmosphere," he wrote, "decreases rapidly and the 'fall-out,' or settling of airborne radioactive particles, is hastened by rain or snow. The latter factor has caused upstate New York areas such as Rochester and Buffalo to be called 'radiation sewers'" (Laurence; see also "Bomb Tests").

In May 1953, Utah stockmen blamed the Nevada tests for the unexpected deaths of more than one thousand ewes and lambs. The AEC investigated the complaints and assured the stockmen that, although they did not know what was responsible for the deaths, it was certainly not atomic tests ("A.E.C. Denies"). A rancher and the wife of another rancher filed suit, claiming they had been injured by the same tests. The woman charged that "radioactive dust from the blasts had caused her hair to fall out, her skin and fingernails to peel off, and gave her recurrent nausea." The man complained of losing all his body hair ("2 Sue").

Although these events received national publicity, it was not until the "Bravo" test of the U.S. superbomb in the Pacific that the scope of the danger of fallout became widely known. Weather forecasts for this 1 March 1954 explosion were wrong, and the fallout was blown in an unexpected direction. Rep. Chet Holifeld (Dem., CA), a member of the Joint Atomic Energy Committee, later characterized the test as "out of control," a charge that the chairman of the AEC denied. This denial was little consolation to the 28 Americans and 236 natives of the Marshall Islands who were exposed to radioactive fallout. Fortunately they were quickly decontaminated and relocated to a safe area ("H-Bomb"; "264 Exposed"; "Big Delay"). The fishermen on a Japanese tuna boat wandering near the Bikini test area were not so lucky.

The twenty-three seamen on the *Lucky Dragon* had no knowledge that a test was about to take place, but the distant, brilliant light in the sky reminded them of stories of Hiroshima and Nagasaki. Nevertheless, they did not recognize the subsequent four-hour snow of strange whitish dust upon their vessel as a special threat. When they soon became ill and began to lose their hair, however, they became alarmed and turned homeward. It was two weeks before they reached Japan, and more days passed before the nature of their illness was discovered. During this period, they worked, ate, and slept in the midst of the fallout dust. After months of illness, most of the men recovered, but Aikichi Kuboyama, the radioman, died on 23 September (Lapp, *Voyage*).

The tragedy was compounded by its impact upon the fishing industry. Many of the fish brought back in the *Lucky Dragon* were found to be contaminated but not until after they had been sold. Radioactive fish were also discovered on other tuna boats, creating near panic in a nation dependent upon the sea for protein. One boat in eight returned with contaminated fish as ocean currents spread radiation from the Bikini test through the Pacific. The national consumption of fish and fish prices plummeted, and the industry suffered terribly (Lapp, *Voyage* 88–100, 177–78; Passin). All of these events were followed closely by the world press.

The bomb that dropped fallout upon the *Lucky Dragon* was very dirty, much more so than one would expect in theory from a hydrogen bomb. It was the first of a new kind of device that used inexpensive uranium 238 in massive quantities. Its deadly fission products and other debris were injected into the upper atmosphere by the blast to circle the globe. Independent scientists identified the nature of the bomb soon after the test and also discovered the presence of strontium 90, a particularly dangerous and long-lasting radioactive isotope (Lapp, *Voyage* 131, 148–56; Rotblat). This was not officially announced, however, until a June 1955 speech delivered by Commissioner Willard Libby of the AEC. He added the reassuring suggestion that, after a nuclear war, fallout could be removed from cities with "ingenious devices such as street sweepers, in which the driver sits on a bag of sand or a thick metal slab to protect him from radiation" (Leviero 32).

The public was now less willing blindly to accept statements like this. The information and apprehensions originally shared by a few scientists were finding their way into the popular press and everyday conversation. Americans became increasingly alarmed when they discovered that their own food was contaminated.

Strontium 90

Oh where, oh where has the fallout gone,
Oh where can the poison be,
Why right in the milk and the other things
That the milkman brings to me.

Sen. George Aiken (Rep., VT) was displeased with this and other songs sung by "certain pacifist groups." In 1962, he asked a congressional hearing witness whether he did not think "it was a great calamity that the critics of the use of milk and other dairy products did not advise the Maker before He set up the original milk program?" (United States, *Radiation Standards* 86–87, 94). The senator's pique was prompted by the universal presence of strontium 90 in milk products, the resulting public anxiety regarding their wholesomeness, and the tremendous emotional leverage that the fear of radioactive milk gave the opponents of nuclear weapons.

A radioactive isotope, strontium 90 (Sr-90) has a half-life of twenty-eight years, making it a long-lasting component of fallout. Soon after World War II, the AEC recognized that Sr-90, which is chemically similar to calcium, can accumulate in bones and possibly lead to cancer ("AEC Fifth Semiannual Report" 119). In August 1953, its presence in animal bones, milk, and soil was first confirmed by the Lamont Geological Observatory. Lamont established a worldwide network for sampling human bone, and within a few years found Sr-90 present in "all human beings, regardless of age or geographic locations" (Kulp et al. 219). Sr-90 found its way into humans via the ecological food chain, as fallout in the soil was picked up by plants, further concentrated in herbivorous animals, and eventually consumed by humans.

The news that Sr-90 was a dangerous component of fallout received wide publicity in 1954 when Japanese scientists discovered that it was a part of the dust sampled from the *Lucky Dragon*. The new superbombs created Sr-90 in far greater quantities than did the old A-bombs. Public concern increased as the 1950s progressed, the bomb tests continued, radiation levels rose, and the issue received a great deal of press attention (Lapp, "Strontium Limits"). In 1956, for example, *Newsweek* announced: "The testing of hydrogen bombs may have *already* propelled enough strontium 90 . . . into the stratosphere to doom countless of the world's children to inescapable and incurable cancer" ("Danger" 88). The magazine characterized Sr-90 as "the invisible bone-hitting particles" that "can never be removed" (88). The federal government established an elaborate system to monitor food and water for Sr-90 and other radioisotopes (Terrill). In addition, there were a number of private research projects that added to knowledge of this pollutant. Some were also designed to increase public awareness of the hazard.

The Consumer's Union, for example, conducted a major national study of Sr-90 concentrations in milk—a highly emotional topic because of the importance of milk in the diet of growing children. Sr-90 was found in a variety of foods in addition to milk, so the organization also conducted an annual study of the Sr-90 levels in the total diet, based upon typical menus of citizens living in a number of cities throughout the United States. The results of these studies were published in a series of articles in *Consumer Reports,* which had a readership in the millions ("Milk"; "Fallout in Our Milk"; "Strontium-90"; "Follow-up Study"; "Fallout and the U.S. Diet"; Michelson).

Another study, the Baby Tooth Survey, was a particularly imaginative combination of research and public education. In 1958, Herman M. Kalckar proposed an international study of the concentration of radioisotopes in baby teeth. The special advantages of baby teeth were that their age could be precisely established; they could, unlike bones, be collected as they were shed without injury to donors; and they were readily available, ensuring a large and continuous supply. Although a coordinated international program was never established, a number of smaller projects were eventually conducted around the world, beginning in St. Louis, Missouri.[4]

In 1958, the newly created Greater St. Louis Citizens' Committee for Nuclear Information decided to undertake a survey of Sr-90 in the teeth of children. The survey started in earnest at the beginning of 1959. To prevent unreasonable public fears as a result of the study itself, and because they would require the assistance of the citizenry, the committee initiated a public-education campaign and successfully gained the support of schools, medical institutions, libraries, Scout groups, and other community organizations. Initially they received teeth at the rate of one thousand each month, but the collection rate rapidly increased. By 1961, one million tooth survey forms had been distributed, and teeth were being received at the rate of 750 each week; a total of more than 67,500 by the end of the year. Nearly 10 percent of these teeth were coming from outside the St. Louis region. They received nearly 160,000 teeth by the end of 1964. Each child was given an I GAVE MY TOOTH TO SCIENCE button ("Baby Tooth Survey Launched"; Reiss 1169–70; "Baby Tooth Survey—First Results"; Logan).

The study became a model for similar projects around the world. Over the years, the scope of the committee broadened. The organization had a strong biological and ecological perspective, and its interests expanded from fallout to wider environmental problems. The name of its newsletter was changed to *Scientist and Citizen,* which eventually metamorphosed into *Environment.* What began as a mimeographed newsletter about fallout had turned into one of the nation's major sources of environmental information; and one of the committee's founders and best-known members, Barry Commoner, had achieved national prominence in the environmental movement (Logan 39; Fleming 40–46).

These and other studies, and the wide publicity they received, brought the issue of radioactive fallout very close to home. No longer was fallout a problem limited to a few Japanese fishermen or western ranchers. People around the nation knew that invisible radioactive material was in the air they breathed and lodged within their own and their children's bones. In learning about this hazard they also learned about the ecological food chain, the biological concentration of these materials, and the cancer and other radiation-induced effects that might strike them in future years.

On the Beach

Public anxiety about the effects of nuclear tests and atomic radiation was expressed in a series of science fiction motion pictures that were long on fiction and short on science. These 1950s masterpieces of the cinematic art included *The Beast from 20,000 Fathoms* (1953; resurrected dinosaur), *Them!* (1954; giant ants), *Tarantula* (1955; giant spider), and *The Incredible Shrinking Man* (1957; tiny man) (see Wright). Rather than representing true science fiction, these films were a modern version of Gothic horror. A literary or cinematic journey from the world of normal, everyday experience to one of fantasy and terror requires some device to encourage belief. Radioactive fallout provided such a device, an excuse for conjuring up demons in the form of mutants, monsters, and nature run amuck. The fall of radiation had become the modern equivalent of the fall of darkness and the stroke of midnight.

Beginning with the 1951 film *Five,* there also arose a new film genre examining the theme of survival after World War III. Other films of this sort included *The World, the Flesh, and the Devil* (1959); *On the Beach* (1959); and *Panic in Year Zero* (1962). Each considered the plight of the survivors of nuclear war: facing one's certain death as radiation spread across the earth; being the last human beings on earth and bearing the responsibility for the future of the species; and surviving in the face of overwhelming disaster and the collapse of social order. Although most of these films were not of the highest quality, they presented to millions of people a terrifying image of the future and expressed the anxieties of their society (Shaheen; Stark).

On the Beach was an exception to the rule. It was a high-budget, prestige film designed to attract international attention to the issues of nuclear war and fallout. The novel, written by the well-known author Nevil Shute, was published in 1957 and became a bestseller with more than two million copies in print by 1960 ("Last Tale"). It portrays the despair and resignation of the citizens of Australia following a 1962 nuclear war in the northern hemisphere. They have to wait over a year for the radioactive air mass of the northern hemisphere to mix sufficiently with the southern air mass to bring their certain death, more than enough time to consider what lies ahead and for each to find his or her own way of coming to terms with the inevitable.

Many reviewers found it difficult to accept the calmness with which Shute's characters face their doom. The Australians go about the business of their lives, adjust to the shortages of supplies, and consider whether or not they will take the government-issued suicide pills that promise a quick death as an alternative to slow radiation poisoning. The trout season is opened early because few will be alive by the time the traditional date arrives. "But there was no orgy of immorality, no riots and looting of the haves by the have nots, no mass religious revival," wrote one reviewer with some disappointment (Prescott). Another, however, wrote, "[I]f this thriller is ever televised, there may be a wilder stampede than Orson Welles wrought two decades ago with his Martians" (Sykes 4).

The film version of *On the Beach,* produced and directed by Stanley Kramer, was released by United Artists two years after the book's publication. With over four million dollars invested, a large figure at that time, Kramer had the difficult task of making a box-office success out of a movie about a terribly depressing subject. He hired big-name stars: Gregory Peck as the American submarine commander, Dwight Towers; Ava Gardner as the less-than-glamorous alcoholic, Moira Davidson; and dancer Fred Astaire as the physicist, Julian Osborne. Casting Gardner and Astaire against type helped attract attention. The promotion of the film emphasized its relevance to major issues of the time—nuclear fallout and the survival of humanity in a nuclear age. This was, as *Variety* noted, "part of United Artists' campaign to make the film what's termed 'a status symbol,' meaning something to be seen despite its grim nature" ("On the Beach"). On 17 December 1959, the film premiered in eighteen cities around the world, with versions in eight languages. Gregory Peck and one thousand others attended the premier in Moscow. With this kind of promotion, it is not surprising that *On the Beach* was a major success. Two months after its release, it was still on top ("Not with a Bang"; "Topics"; Crowther; "Gregory Peck"; "National Box Office").

Perhaps the most moving of the film's scenes are those of the dead cities of the U.S. West Coast. Towers and his crew are sent to investigate strange radio signals coming from the area, hoping they are a sign that some human life remains. Raising the periscope to examine the coast, they find cities devoid of life. The images of San Francisco, its streets empty and without movement, are haunting. The final disappointment comes when they discover that the radio signals, which they have traveled halfway around the world to investigate, are created as a window shade randomly flapping in the breeze jiggles a Coke bottle against a telegraph key.

The film's final scenes of Melbourne's vacant, lifeless streets recall Dwight Towers's thoughts in the novel as he drives through the city.

> Very soon, perhaps in a month's time, there would be no one here, no living creatures but the cats and dogs that had been granted a short reprieve. Soon they too would be gone; summers and winters would pass by and these houses and these streets would know them. . . . The human race was to be wiped out and the world made clean again for wiser occupants without undue delay. (Shute 276–77)

In addition to widespread and strong praise for the film, there were notes of criticism. Some reviewers leveled the same charge as had been directed at the book, arguing that the characters accepted their fate too calmly. Others complained that the film did not show the violence of the war or the physical agony of its victims. A *Time* reviewer wrote that the film "turns out to be a sentimental sort of a radiation romance, in which the customers are considerately spared any scenes of realistic horror" ("New Picture"). Lodging a different criticism, New York's Gov. Nelson Rockefeller feared the film might diminish the nation's "will to resist. . . . [S]ome of my kids saw the picture and came away with the feeling of 'what's the use?'" (Illson 42).

After a decade of preparation, the American public was ready to believe what *On the Beach* had to say. People understood that fallout can circle the globe and that this invisible poison, which they were unable to detect with their own senses, could threaten their lives and future. At a congressional hearing in mid-1961, Herman Kahn spoke of the scientists in the 1950s who did not believe nuclear war was survivable. "In other words," he said, "the belief in the 'end of history' was an expert's belief, rather than a layman's belief. In fact, if the layman had been told fully and frankly what the experts believed, he would have been horrified. . . . The picture and book, 'On the Beach,' reflected these views" (United States, *Civil Defense* 178–79). The "end of history," however, was no longer a concept known only to experts.

Seeking Shelter

Americans did not accept the "end of history" passively. The mid- and late 1950s witnessed growing public and congressional interest in fallout shelters. Gov. Rockefeller was a vocal advocate; and in the spring of 1960, he announced plans to build one in the basement of his New York Fifth Ave. apartment building. He made a special effort to influence the new president, John F. Kennedy. Given the well-known hazards of fallout and nuclear war, it was difficult for the president not to take steps to protect the population from this potentially disastrous threat. In a special message to Congress on 25 May 1961, he announced a major step-up in the nation's civil-defense program (Simpson; Illson; *John F. Kennedy* 403).

In June, Kennedy met with Premier Nikita Khrushchev of the U.S.S.R., who told him of the Soviets' intention to end the West's access to Berlin. In response, Kennedy made a radio and television report to the nation announcing an increase of $207 million above the $104 million already appropriated for civil defense—a total of five times the previous year's funding. This was only one part of a major mobilization of U.S. defense in preparation for the likelihood that the Soviet Union would sign a separate peace treaty with East Germany, thus isolating West Berlin. This was a grim message, in which he raised the specter of nuclear war with the U.S.S.R. The president's speech prompted an outpouring of national concern. In July, the Office of Civil and Defense Mobilization received 16,994 inquiries from the public, with a major increase following the speech—5,382 letters on 1 August alone (*North Atlantic* 42; *John F. Kennedy* 533–40; Kaplan 309; United States, *Civil Defense* 68).

On the night of 13 August, East Germany began constructing the Berlin Wall. The Soviet Union resumed testing nuclear weapons on 31 August. Since the end of 1958, the two nations had tacitly agreed to suspend nuclear testing; and between that time and August 1961, neither country had conducted tests. By the end of 1961, though, the Soviets had detonated more than thirty devices. The United States reestablished its own testing program and by the end of 1962 had detonated nearly ninety devices, compared to about forty of the Soviets' in the same year. The world of 1962 witnessed the largest annual number of nuclear explosions in history (*North Atlantic* 43; *SIPRI: 1968/69* 242; *SIPRI: 1983* 100). The background radiation level, which had dropped since 1958, again began to climb as nuclear debris was injected into the atmosphere.

The hostilities between the two nations reached a peak in October 1962, when Kennedy decided to confront the Soviets over their attempt to base nuclear missiles in Cuba. Five years after the publication of *On the Beach*, in the year of the novel's fictional holocaust, the world held its breath as the two superpowers poised on the edge of a terrifyingly real nuclear war.

Through the autumn of 1961, the administration had continued to promote the creation of public and private fallout shelters. *Life* magazine published a major article on fallout shelters in September, complete with an introductory letter from the president. "Nuclear weapons and the possibility of nuclear war," he wrote, "are facts of life we cannot ignore today. . . . The ability to survive coupled with the will to do so therefore are essential to our country." The article claimed that "97 out of 100 people can be saved" and provided diagrams of home shelters (including one soon to be available from Sears, Roebuck and Co. for seven hundred dollars) and tips on shelter living ("Fallout Shelters"). In December, the Defense Department published a brochure promoting home shelters and other forms of fallout protection. Twenty-five million copies were distributed free from post offices and civil-defense offices throughout the nation (Kaplan 313).

The country was swept up into "shelter mania" as citizens with the means constructed fallout shelters in their basements and backyards. Entrepreneurs marketed kits of food and survival equipment for the well-outfitted shelter, and clothing stores catered to the special needs of doomsday. One Manhattan dress shop recommended "gay slacks and dress with a cape that could double as an extra blanket" ("Fallout Shelters"; "Survival" 19).

Shelter mania showed its dark side as citizens armed their home shelters to fight off neighbors who, in the event of a war, might want to share their limited space and provisions. A Nevada civil-defense official announced that it might become necessary to rely on vigilantes to defend his state from World War III Californian refugees ("Fall-Out Shelters Speeded"). "There is evidence that the Administration policies, which seem to emphasize an every-man-for-himself approach," wrote *Newsweek,* "have succeeded in bringing out the worst side of human nature. Some citizens are behaving as if they were cavemen already" ("Survival").

Criticism of the program grew. In November, Kennedy's adviser Arthur Schlesinger warned the president, "Everywhere the shelter program seems to be emerging as the chief issue of domestic concern—and as one surrounded by an alarming amount of bewilderment, confusion and, in some cases (both pro and con) of near-hysteria" (Kaplan 312). *Newsweek* and *Consumer Reports* pointed out that the administration's program did not provide protection from blast, heat, or firestorm, and did not provide for dispersing targets ("Survival"; "Fallout Shelter" 14). In December, the American Medical Association urged the nation to "stop worrying about radioactive fall-out and concentrate on getting ready for Christmas." It went on to say, "There really isn't very much that us average folks can do about it anyway"; and, "If there is radioactive fall-out in the air, we'll get some of it, and there's nothing we can do about it" ("Stop").

As the mania abated and cooler heads prevailed, Congress pared the president's civil-defense-budget request for fiscal year 1962–63 from $695 million down to $80 million. Steps were taken to reduce the hazards of nuclear weapons with the signing in June 1963 of a treaty to install a "hot line" between Moscow and Washington and in August 1963 of the Limited Test Ban Treaty to halt above-ground testing (Kaplan 314; *North Atlantic* 45). The U.S. government would continue to support fallout shelters, but never again would the public display the kind of obsession that had characterized this period. Kennedy's civil-defense program left a lasting impression upon the nation. The "end of history," nuclear war, and radioactive fallout were no longer simply items of uncomfortable conversation. They were threats against which individual citizens had physically prepared. Worse still, the actions that their government urged were not designed for prevention. Instead, they were based upon accepting and accommodating to this overshadowing doom. In the following years, the ubiquitous fallout-shelter sign and its radiation symbol became part of the landscape as it graced schools, public buildings, subway tunnels, and many privately owned structures. It became a reminder of a terrifying, inescapable threat.

Chemical Fallout

Silent Spring was published on 27 September 1962—one month before the Cuban missile crisis and one year before the signing of the Limited Test Ban Treaty; almost three years after the release of the film version of *On the Beach* and two years before the release of Stanley Kubrick's *Dr. Strangelove: Or, How I Learned to Stop Worrying and Love the Bomb*. The nation was steeped in years of debate about nuclear weapons and fallout, which served as a point of reference to help people understand the hazards of pesticides and as a fearful symbol to motivate action.

The environmental and health hazards of radioactive materials were on Rachel Carson's mind as she wrote the book. In the summer of 1960, while deeply involved in writing *Silent Spring,* she also worked on a revised edition of *The Sea Around Us.* In a new preface, she wrote about the impact of fallout and of the ocean disposal of nuclear wastes upon the marine environment. She described how marine organisms can concentrate radioisotopes and wrote, "By such a process tuna over an area of a million square miles surrounding the Bikini bomb test developed a degree of radioactivity enormously higher than that of the sea water." In creating these materials, she warned, we must face the question of whether we "can dispose of these lethal substances without rendering the earth uninhabitable" (xi–xiii).

It is no accident, then, that the first pollutant Carson mentioned by name in *Silent Spring* was not a pesticide but strontium 90. Well known to the American public, Sr-90 was a tool to help her explain the properties of pesticides. Early in *Silent Spring* she wrote:

> Strontium 90, released through nuclear explosions into the air, comes to earth in rain or drifts down as fallout, lodges in soil, enters into the grass or corn or wheat grown there, and in time takes up its abode in the bones of a human being, there to remain until his death. *Similarly,* chemicals sprayed on croplands or forests or gardens lie long in soil, entering in a chain of poisoning and death. (6, emphasis added)

Although this is the book's first reference to a specific pollutant, it is not its first allusion to fallout. The opening chapter, "A Fable for Tomorrow," paints a picture of a lovely rural midwestern town struck by a mysterious blight. People, animals, fish, and birds sicken and many die. Roadside vegetation withers. What has happened to this town, now lifeless and without even the song of birds? In the nooks and crannies of the town's buildings, one can find a white powder that "had fallen like snow upon the roofs and the lawns, the fields and streams" a few weeks before. "No witchcraft," she wrote, "no enemy action had silenced the rebirth of new life in this stricken world. The people had done it themselves" (3). This fall of pesticides upon the town conjures up the specter of radioactive fallout—a specter created intentionally by the author. In an early draft, Carson had written that the powder reminded the townspeople of the dust that fell upon the *Lucky Dragon*. She had also written that visitors to the town wondered if perhaps the wind had carried fallout from a bomb test and dropped it on the town (Rachel Carson Papers, "Chapter 1"). Not only does this chapter present a frightening description of potential pesticide hazards, it evokes the image of a town dying from nuclear fallout. On an even more subtle level, it recalls the images of lifeless American cities shown so graphically less than three years before in *On the Beach*.

Elsewhere in her book, Carson made an even more direct comparison between fallout and pesticides. Writing of a Swedish farmer who had died of pesticide poisoning and recalling the unfortunate radioman of the *Lucky Dragon,* she wrote, "Like Kuboyama, the farmer had been a healthy man, gleaning his living from the land as Kuboyama had taken his from the sea. For each man a poison drifting out of the sky carried a death sentence. For one, it was radiation-poisoned ash; for the other, chemical dust" (229–30). A few pages later, she wrote, "Certain chemicals, again reminding us of radiation products like Strontium 90, have a peculiar affinity for the bone marrow" (234). She also referred to the leukemia victims of the Hiroshima A-bomb to illustrate a similar hazard from pesticides (226). Other references to radiation are sprinkled throughout the book.

Lois and Louis Darling, *Silent Spring*'s illustrators, also had radiation in mind as they explored ideas for drawings. Their margin notes on a draft manuscript include a mushroom cloud sketch in one place and a note to illustrate the Swedish farmer–*Lucky Dragon* comparison in another (Rachel Carson Papers, "Typescript"). I have found no evidence that Rachel Carson directly suggested either of these possibilities to the Darlings. Although neither of these ideas found their way into the final book, they demonstrate the images the book brought to mind.

I am not suggesting that using fallout as an analogy for pesticides was a central part of the design of this very sophisticated book. As a thoughtful person who was aware of the issues of her time, however, it was impossible for Carson not to have been influenced by the decade of public discussion and debate. Both Carson and her editor, Paul Brooks, were well aware of the similarities between the effects of fallout and pesticides. And while, when interviewed nearly twenty-two years after publication of *Silent Spring,* Brooks did not recall that this was a major part of their conversation (Brooks, Personal interview), there is now evidence that he had suggested to Carson that she make the comparison.[5] Carson and her book were products and representatives of their time, as well as shapers of it.

Fallout, one might say, was "in the air" and it is a tribute to Carson's perceptive skill as an author that she was able to recognize and take advantage of the deep-seated cluster of social concerns surrounding it in the public's mind. Not only did she tap into this anxiety and direct it toward pesticides, she also used the public's existing understanding about the hazards of fallout to teach about the similar hazards of chemical poisons. Just as strontium 90 could travel great distances, enter the food chain, and accumulate in human tissue, so too could pesticides. Just as radioactive materials could produce chronic rather than acute poisoning, so too could pesticides. And just as exposure to radiation could produce cancer, birth defects, and mutations, so might pesticides. The public already knew the basic concepts—all it needed was a little reminding.

A distinctive feature of the contemporary environmental movement is a profound and pervasive element of fear. It is a fear that, for good or ill, colors and sometimes distorts virtually every popular analysis of major environmental problems. This is not simply a fear that we will deplete a particular natural resource, lose pristine wilderness, or be poisoned. It is the belief that we may well be facing the "end of history," that we as a species might be doomed. This anxiety burst to the surface with the destruction of Hiroshima and Nagasaki. It is rooted in the omnipresent threat of nuclear destruction.

The generation that promoted Earth Day 1970 grew up in the shadow of nuclear destruction. This threat became a tacit part of the way in which people understood their world. It is no surprise then, that the belief in the imminent end of the earth became integrated with more traditional conservation concerns. This younger generation did not create the anxiety, nor did its elder, Rachel Carson. She did, though, write one of the first and most eloquent books bridging the gap between the environmental movement and this new fearful vision of Armageddon.

Notes

I wish to thank Prof. Allan Krass, Hampshire College, for his helpful comments on a draft of this essay.

1. For the history of *Silent Spring* and the controversy surrounding it, see Linda Lear's biography of Carson; as well as Brooks, *House;* and Graham. See also Ehrlich.

2. For an examination of the history of pesticides and their regulation in the United States, see also Dunlap; Rudd; and Graham.

3. This is not to say that this issue—radioactive fallout and nuclear waste—was never mentioned. A number of writers included them in their lists

of pollutants that were of public concern at the time. What I am proposing here is that the decade of public discussion and anxiety about these pollutants, particularly fallout, that preceded the publication of *Silent Spring* played a special role in preparing the public to accept Rachel Carson's message. It is this point that has largely been overlooked. Fleming has come closest to identifying this special relationship between fallout, pesticides, and the contemporary environmental movement: "Anybody who has been alarmed by atmospheric pollution from nuclear tests could see that [Carson] was talking about other dimensions of the same problem. She, for her part, invoked the menace of strontium 90 as an ominous backdrop to her indictment of DDT" (43). This special relationship, the educational impact of the fallout controversy, and the use of fallout as an analogy of pesticides are the foci of this essay. Dunlap provides the most extensive examination of the ways that bomb tests and fallout prepared the public to question the benefits of technology, including pesticides (74, 102–4, 187; see also Nash 252). Weart provides a detailed examination of the American cultural response to the phenomenon of radioactivity through the twentieth century.

4. The number of Sr-90 citations in the *New York Times Index* rose rapidly in 1957, peaked in 1959, and dropped sharply the next year, never to return to the 1957 level. This suggests that the Consumer's Union, St. Louis, and related studies came late and informed, rather than precipitated, the public discussion.

5. After the earlier version of this essay was first published, Linda Lear found a 29 March 1960 letter from Brooks to Carson in which he suggested that comparing radiation and chemicals would help awaken her readers to the dangers of pesticides (see Lear 374–75).

Works Cited

"A.E.C. Denies Rays Killed Utah Sheep." *New York Times* 17 Jan. 1954, sec. 1: 46.

"AEC Fifth Semiannual Report: Part II." *Bulletin of the Atomic Scientists* 5 (Apr. 1949): 114–25.

"Baby Tooth Survey—First Results." *Nuclear Information* Nov. 1961: 1.

"Baby Tooth Survey Launched in Search of Data on Strontium 90." *Nuclear Information* Dec. 1958: 1–5.

"Big Delay in Radiation." *New York Times* 13 Mar. 1954, 32.

"Bomb Tests Are Cleared as Cause of Rain in East." *New York Times* 22 May 1953, 3.

Brooks, Paul. *The House of Life: Rachel Carson at Work*. Boston: Houghton, 1972.

———. Personal interview. 27 Aug. 1984.

Carson, Rachel. Papers. "Chapter 1," Folder 39-1. Yale Collection of American Literature. Beinecke Rare Book and Manuscript Library, Yale University.

———. *The Sea Around Us*. Rev. ed. New York: Oxford UP, 1961.

———. *Silent Spring*. Boston: Houghton, 1962.

Crowther, Bosley. "Screen: 'On the Beach.'" *New York Times* 18 Dec. 1959, 34.

"Danger—Strontium 90." *Newsweek* 12 Nov. 1956: 88, 90.

Downs, Robert B. *Books That Changed America*. New York: Macmillan, 1970.

Dunlap, Thomas R. *DDT: Scientists, Citizens, and Public Policy*. Princeton: Princeton UP, 1981.

Eherhart, Sylvia. "How the American People Feel about the Atomic Bomb." *Bulletin of the Atomic Scientists* 3 (June 1947): 146.

Ehrlich, Paul R. "Paul Ehrlich Reconsiders *Silent Spring*." *Bulletin of the Atomic Scientists* 35 (Oct. 1979): 34–36.

"Fallout and the U.S. Diet." *Consumer Reports* Mar. 1962: 139–43.

"Fallout in Our Milk." *Consumer Reports* Feb. 1960: 64–70.

"The Fallout Shelter." *Consumer Reports* Jan. 1962: 8–14.

"Fallout Shelters." *Life* 15 Sept. 1961: 95–108.

"Fall-Out Shelters Speeded by Hundreds in Suburbs." *New York Times* 3 Oct. 1961, 41.

Fleming, Donald. "Roots of the New Conservation Movement." *Perspectives in American History* 6 (1972): 7–91.

"A Follow-up Study on Strontium-90 in the Total Diet." *Consumer Reports* Oct. 1961: 547–49.

Fowler, John M. *Fallout*. New York: Basic, 1960.

Graham, Frank, Jr. *Since Silent Spring*. Boston: Houghton, 1970.

"Gregory Peck Sees His Film in Moscow." *New York Times* 18 Dec. 1959, 34.

"The H-Bomb." *New York Times* 4 Apr. 1954, sec. 4: 1.

Illson, Murray. "Rockefeller to Build Apartment Shelter." *New York Times* 5 Mar. 1960, 1+.

"Immortal Nominations." *New York Times Book Review* 3 June 1979: 12–13, 51.

John F. Kennedy: Containing the Public Messages, Speeches, and Statements of the President, January 20 to December 31, 1961. Public Papers of the Presidents of the United States. Washington: GPO, 1962.

Kalckar, Herman M. "An International Milk Teeth Radiation Census." *Nature* 182 (1958): 283–84.

Kaplan, Fred M. *Wizards of Armageddon*. New York: Simon, 1983.

Kulp, J. Laurence, Walter R. Eckelmann, Arthur R. Schulert. "Strontium-90 in Man." *Science* 125 (1957): 219–25.

Lapp, Ralph H. "Strontium Limits in Peace and War." *Bulletin of the Atomic Scientists* 12 (Oct. 1956): 287–89, 320.

———. *The Voyage of the Lucky Dragon.* New York: Harper, 1958.

"Last Tale of a Skilled Story Teller." *Saturday Review* 2 Apr. 1960: 16.

Laurence, William L. "Vast Atom Strides Expected in Tests." *New York Times* 15 Mar. 1953, sec. 1: 83.

Lear, Linda. *Rachel Carson: Witness for Nature.* New York: Holt, 1997.

Leviero, Anthony. "Cheap H-Bomb Is Now Possible." *New York Times* 12 June 1955, sec. 1: 1+.

Logan, Yvonne. "The Story of the Baby Tooth Survey." *Scientist and Citizen* Sept.–Oct. 1964: 38–39.

Michelson, Irving. "Monitoring Fallout from Nuclear Weapons Tests." United States, *Radiation Standards* 227.

"The Milk All of Us Drink—And Fallout." *Consumer Reports* Mar. 1959: 102–3.

Nash, Roderick. *Wilderness and the American Mind.* Rev. ed. New Haven: Yale UP, 1973.

"National Box Office Survey." *Variety* 2 Mar. 1960: 3.

"New Picture." *Time* 28 Dec. 1959: 44.

The North Atlantic Treaty Organization: Facts and Figures. 10th ed. Brussels: NATO Info. Service, 1981.

"Not with a Bang or a Whimper." *Saturday Review* 24 Oct. 1959: 32–33.

"On the Beach." *Variety* 2 Dec. 1959: 6.

Passin, Herbert. "Japan and the H-Bomb." *Bulletin of the Atomic Scientists* 11 (Oct. 1955): 289–92.

Prescott, Orville. "Books of the Times." *New York Times* 24 July 1957, 23.

———. "Typescript (carbon) with ms annotations by the illustrators," unnumbered box marked "The Sea Around Us/Silent Spring." Yale Collection of American Literature. Beinecke Rare Book and Manuscript Library, Yale University.

Reiss, Louise Zibold. "Strontium-90 Absorption by Deciduous Teeth," *Science* 134 (1961): 1669–73.

Rotblat, J. "The Hydrogen-Uranium Bomb." *Bulletin of the Atomic Scientists* 9 (May 1955): 171–72, 177.

Rudd, Robert L. *Pesticides and the Living Landscape.* Madison, U of Wisconsin P, 1964.

Shaheen, Jack G., ed. *Nuclear War Films.* Carbondale: Southern Illinois UP, 1978.

Shute, Nevil. *On the Beach.* New York: Morrow, 1957.

Simpson, Mary M. "A Long Hard Look at Civil Defense." *Bulletin of the Atomic Scientists* 12 (Nov. 1956): 343–48.

SIPRI Yearbook of World Armaments and Disarmament: 1983. London: Taylor, 1983.

SIPRI Yearbook of World Armaments and Disarmament: 1968/69. Stockholm: Almquist, 1970.

Stark, Steven. "Nuclear Holocaust Is Big at Box Office." *Boston Globe* 4 Aug. 1985, A17+.

"Stop Fall-Out Worry, AMA Urges Public." *New York Times* 15 Dec. 1961, 26.

"Strontium-90 in the Total Diet." *Consumer Reports* June 1960: 289–93.

"Survival: Are Shelters the Answer?" *Newsweek* 6 Nov. 1961: 19.

Sykes, Gerald. "Supine Surrender." *New York Times Book Review* 20 July 1957: 4+.

Taussig, Helen B. "The Thalidomide Syndrome." *Scientific American* (Feb. 1962): 29–35.

Terrill, James G., Jr. "Monitoring—Surveillance Activities in the United States Public Health Service Surveillance Programs." United States, *Radiation Standards* 179–220.

"Topics: 'On the Beach.'" *New York Times* 17 Dec. 1959, 36.

"264 Exposed to Atom Radiation after Nuclear Blast in Pacific." *New York Times* 12 Mar. 1954, 1+.

"2 Sue for $200,000 in Atomic Radiation." *New York Times* 4 May 1954, 12.

United States. Cong. House of Representatives. Hearings before a Subcommittee of the Committee on Government Operations. *Civil Defense—1961.* 87th Cong., 1st sess. Washington: GPO, 1961.

United States. Cong. Senate. Subcommittee on Reorganization and International Organizations of the Committee on Government Operations. *Interagency Coordination in Environmental Hazards (Pesticides).* 88th Cong., 1st sess. 1964, pt. 1. Washington: GPO, 1964.

United States. Cong. Subcommittee on Research, Development, and Radiation of the Joint Committee on Atomic Energy. *Radiation Standards, Including Fallout.* 87th Cong., 2nd sess. 1962.

Weart, Spencer R. *Nuclear Fear: A History of Images.* Cambridge: Harvard UP, 1988.

Whitaker, Adelynne Hiller. "A History of Federal Pesticide Regulation in the United States to 1947." Diss. Emory U, 1974.

Whorton, James. *Before Silent Spring: Pesticides and Public Health in Pre-DDT America.* Princeton: Princeton UP, 1974.

Wright, Gene. *The Science Fiction Image.* New York: Facts on File, 1983.

3

CHRISTINE ORAVEC

An Inventional Archaeology of "A Fable for Tomorrow"

❧ The central significance of Rachel Carson's *Silent Spring* for the modern environmental movement is nearly undisputed (Brooks 227; Graham x, xii; Hynes 9; Killingsworth and Palmer 9–16). Similarly, one might argue that the first chapter of *Silent Spring,* entitled "A Fable for Tomorrow," is at the very center of the controversies surrounding the book. Because it combined the two incompatible genres of mythic narrative and scientific fact, this chapter contributed more than any other to Carson's being variously labeled a hysterical alarmist and a poetic amateur; one critic even called the chapter "science fiction" (Graham 64–65; Hynes 12–13). Yet this chapter also considerably influenced the controversial genre of apocalyptic

An earlier version of this essay appeared in the 1995 *Proceedings of the Conference on Communication and Our Environment.* Edited by David B. Sachsman, Kandice Salomone, and Susan Senecah. Used by permission of the author.

environmental writing by inspiring both admiring imitation and cynical parody (Brooks 294–95, 297; Hynes 115–18; Killingsworth and Palmer 11). And it is the part of the book that readers remember most. In fact, "A Fable for Tomorrow" has gained a reputation as a rhetorical bombshell that landed in just the right place at just the right time.

In addition to determining the general effects of discourse, however, communication scholars tend to make more technical inquiries. Among those inquiries are questions about the origins of discourse, or what the ancient rhetoricians labeled the canon of invention. In the case of "A Fable for Tomorrow," those questions might proceed along the following lines: How did Carson construct this piece of effective discourse? How did she generate the idea for the chapter? What were the serious choices about strategy, technique, and tone she made before the final version saw print? What did she decide to eliminate from previous versions, and why? In other words, out of the resources available, how did Carson choose the most effective means of persuasion in this case? Each of these questions can be answered through an examination of the author's inventional processes.

Inventional processes are not easy to examine directly since the evidence is often elusive and immaterial. We are not often invited to sessions in which rhetoric is planned before there is any evidence of its possible effectiveness; we are seldom one of the inner circle that implements the selection of materials, the construction of arguments, or the tone; and infrequently do we overhear the endless stream of talk that articulates the rationale behind the choice of particular words and phrases. Our only sources of evidence for the processes of invention are the traces that are left in the documents associated with the process, including manuscript drafts, source materials, and written memoirs. The process of unearthing and reassembling these fragments is multilayered and inferential; and in the end, we must construct yet another narrative that makes sense of the material remains.

Of all of the human arts and sciences, the one that this process of unearthing and reconstructing most resembles is that of archaeology. And as everyone knows, archaeologists are particularly attracted to those who are collectors themselves. In the case of Rachel Carson, we have what amounts to an inventional archaeologist's dream: an author self-conscious enough about her art that, by the time she began *Silent Spring,* she collected all of her preliminary materials as they came to her, arranged her successive drafts in sequence, and never, it seems, threw a scrap of paper away. By examining the materials Carson collected in the course of writing "A Fable for Tomorrow," we have an opportunity to recreate the persuasive directions taken by an early, leading environmental writer in the course of creating her most well known text. We can do an inventional archaeology upon an environmental discourse that really has made a difference.

By using the term *inventional archaeology,* I wish both to associate my methodology with and dissociate it from Michel Foucault's famous concept. To the degree that my procedure engages the problem of the constitution of historical series based upon fragmentary and discontinuous traces of discourse, I wish to claim affinity with Foucault's methodological concerns. But I make no claims to be describing entire epistemes or discursive formations by virtue of this activity. My purpose is much more immediate; it is to focus upon Carson's rhetorical problem and its practical solution. To borrow Foucault's terms, I see my story of the various drafts of "A Fable for Tomorrow" *neither* as a document (a mere record of intention) *nor* as a monument (a text in itself, standing apart from its surrounding context) but as a powerful and functional rhetorical construct, profoundly implicated in its historical place and time. Perhaps this particularity and situatedness indicates an episteme; perhaps not.

In any case, this project does assume the importance of the origin and influence of a discourse, which is not necessarily Foucault's purpose, particularly in *The Archaeology of Knowledge* (25–26).

This study is a reconstruction of the inventional processes involved in the writing of "A Fable for Tomorrow," from the first gathering of source materials to the last corrections made upon the galley proofs. To support the reconstruction, I use the best source available, the manuscripts archived in the Rachel Carson Papers of the Beinecke Rare Book and Manuscript Library at Yale University. By examining these folders in roughly chronological order, I aim to uncover the content, alternatives, and outcomes of Carson's rhetorical choices and to generate a narrative that reconstructs the process by which Carson invented this most important chapter. More specifically, I adopt the pretense that I am looking at each folder in sequence without anticipating the content of the succeeding folder. In doing this, I wish to imitate the way rhetorical scholars may experience the process of uncovering the source of a rhetor's strategic moves as though they were eavesdropping upon an evolving inventional conversation.[1] Such an examination of Carson's manuscripts can help us recreate the roots of the recent past in the environmental movement and also help us reflect about our own activity, both as critics and as rhetoricians.

"A Fable for Tomorrow": A Story of Invention

Inventional strategies are often stimulated and shaped by particular rhetorical problems. Digging around in Carson's preparatory materials highlights the kinds of problems she faced and exposes the characteristic and effective ways she responded to these problems. Carson's main problem was to make real and immediate a general threat to health and life from chemical pesticides—a threat that seemed very remote to her readers. To solve this problem, Carson fictionalized a body of factual evidence for the threat yet retained the credibility of those facts. In particular, she chose to employ the device of mythic storytelling to engage her readers and heighten their concern.

When writing about Carson's "mythologizing" of the fictional elements of *Silent Spring,* I am definitely not using the term in its pejorative sense (that is, to claim that myth is "false"), but rather I use it to recapture the connotations of universalization and archetypicality that myths provide. I also do not mean to suggest that Carson, the "real author" in Wayne Booth's phrase, intended to mythologize as a specific and conscious strategy (Booth 86). Instead, I am using "Carson" here to represent the author's choices as represented in the textual outcome, given the range of possibilities revealed in the preceding manuscripts and taking into account such external materials as letters or diaries that might shed light on the process of choice. That is, I use the extant traces of authorial intent to rein in the general domain of possibility and from there project options and choices from the texts of the manuscripts themselves.

"A Fable for Tomorrow" was precisely what Carson provided in this first chapter: a fabulous tale with a moral lesson. The archaeology of this fable, however, reveals that it was constructed, layer by layer, using elements of other discursive genres and modes. Basing the fable upon a foundation of factual reportage, Carson successively employed techniques of realist novelization and archetypal mythologizing, traces of which remain in the final product. Thus, far from generating an example of a single literary form, Carson produced a composite that emerged from several concrete examples and from her own formal experimentation and selection.

Out of the Rachel Carson Papers collection, which is comprised of more than one hundred boxes of materials, I have selected the specific folders that I believe most clearly trace the story of the development of "A Fable for Tomorrow." For purposes of clarity, I have arbitrarily numbered them Folders 1–8 (see the works cited for the corresponding archival index numbers). Looking at each folder in turn reveals how Carson made important strategic decisions at each step in the process. The first folder gives a glimpse of how she gathered her source materials for writing the first chapter of *Silent Spring.*

Folder 1: Source Materials

As I have suggested, one important outcome of "A Fable for Tomorrow" was defining a relationship between the fictional and the factual. But nothing in the folder of source materials suggests that the first chapter will be literary or fictional in the least. These source materials are varied, but all of them are factually based. Among them are clippings about pesticides from the *New York Times* (one entitled "Insect War Brooks No Truce"), articles from the *Saturday Review* and from the newsletter of the Defenders of Wilderness, as well as magazine advertisements from industrial manufacturers.

Apparently, given some incidental references in her notes within Folder 1, Carson had already read several stories about the effects of pesticides on wildlife and people in particular places throughout the country, stories that would play an important part in the invention of "A Fable for Tomorrow." These stories could be found in such publicly available outlets as John K. Terres's article in the *New Republic* entitled "Dynamite in DDT," describing a mass die-off of songbirds in Moscow, Pennsylvania. To quote Terres, after a spraying of DDT on a large oak stand, "[T]he sun arose on a forest of great silence—the silence of total death. Not a bird call broke the ominous quiet" (415).[2]

The notes included in this folder also contain some early outlines. Carson's original conception for the first chapter is a straightforward summary presenting the book's thesis:

that in at least one major area of man's efforts to gain mastery over nature—the reduction of unwanted or 'pest' species—the control operations are themselves dangerously out of control, with the result that normal and necessary relations of living things to each other and to the earth have been destroyed.

Later in these materials, the chapter is given a name, "The Rain of Death," and the summary is considered to "stand for the book in miniature."

This chapter title, however, changed many times in the course of the revision. In the contract with Houghton Mifflin, the book was tentatively called "The Control of Nature" (Lear, *Rachel Carson* 324). But both Carson and Paul Brooks, editor in chief of Houghton Mifflin's general book department, felt this title was too broad. The manuscripts indicate that several titles were considered before *Silent Spring* was agreed upon, the two most prominent of which were "Man's War Against the Earth," and "Man Turns Against the Earth" (Brooks 239; Graham 21). At least one other title, "The War Against Nature" or "At War with Nature," was considered, as indicated in a letter from Carson to her friend Dorothy Freeman. The letter also includes the following thoughts about the beginning of the book: "I told you that a possible opening sentence had drifted to the surface of my mind recently. It was—'This is a book about man's war against nature, and because man is part of nature it is also and inevitably a book about man's war against himself'" (Carson, "To Dorothy Freeman" 380).

Finally, Folder 1 contains a guest editorial by Mark Van Doren for the *Saturday Review* entitled "A Motto for Today." This article seems to suggest that the title "A Fable for Tomorrow" may have been conceived in the context of contemporary doubts about the claims for the supremacy of science, which was the theme of Van Doren's column. But the conception of the book as a whole, including its final title, requires the far more intensive development to be found in the second folder.

Folder 2: Outline of the Book and Introductory Notes

In the second folder, more complete outlines of the book appear. In one outline, the phrase "Silent Spring" first emerges as the title of the first chapter in an eighteen-chapter book. (In a 13 September 1960 letter to Carson, Paul Brooks proposed that the bird chapter [chapter 8] be titled "Silent Spring"; in a 2 December 1960 letter to Carson, Carson's agent, Marie Rodell, recommended this title for the entire book; finally, in a 26 May 1961 letter to Carson, Brooks offered the same recommendation [Lear, *Rachel Carson* 375, 377, 386].) Here, the first chapter is not presented as a summary; rather, it is followed by another chapter heading entitled "Background of the problem" (possibly chapter 2), which might serve as a summary instead. In another outline as well, this first chapter appears to lose its character as a straightforward summary. Rather, it seems to set out a specific theme, focusing upon the enigmatic dichotomy "Life vs. environment."

What *kind* of discourse, then, will chapter 1 become? In subsequent notes, we get a first view of its emerging fictional nature. A sketchy plot is outlined in which a malevolent spell is cast over an unwitting community, highlighting a contrast between "idyllic beauty" and a "nightmare landscape treated with pesticides." Included are pictures of the shore, the forest, orchards, crabs, and children both before and two weeks after a pesticide application. The chapter then would conclude with a series of factual events from around the country, reflecting Carson's research—robins disappearing from the campus of a midwestern university, for example. Carson here appears to be developing a composite, hence fictionalized, portrait of a community but still grounds the composite in factual reports of several communities in the United States that were already experiencing pesticide damage. In the next iteration, however, Carson frees the story line of chapter 1 from its factual underpinnings and develops a novelistic narrative that demonstrates her genius but is also destined to spark controversy.

Folder 3: Various Drafts of Chapter 1

Those familiar with "A Fable for Tomorrow" may be surprised to learn of the degree of fictionalization, indeed almost novelization, the chapter receives in these important drafts. The name of the town, for example, is Green Meadows; it is visited in 1965 (three years after the publication of the book itself) by a young man longing to see the fields, trees, and animal life as he once knew them; he senses that something is subtly wrong; the farmers tell him stories of illness among children; and he notices that the birds are dying. Many of the stories and episodes the young man witnesses are drawn in detail from those events Carson had collected from the factual reports mentioned in the previous folder. In one draft, she underscores the point by stating that every incident she descriptively narrates has occurred somewhere in the United States.

In addition to providing a fictionalized hero in these earlier drafts, Carson employs a fictionalized homodiegetic narrator (presumably the young man himself), an *I* who "relates" or "writes" the "strange and nightmarish events" of what happened to the town of Green Meadows. Eventually, however, throughout several drafts, the young man and even the homodiegetic narrative voice disappear. Instead, a nonfictional, heterodiegetic narrator emerges—a narrator easily identifiable as Carson herself—in the crucial concluding paragraph. It is this paragraph that contains the claim for the factuality of the generalized composite of Green Meadows.

In the later drafts, the description tightens, becomes less detailed and didactic, and the manuscripts decrease in size by half.

Green Meadows retains its name, but the title "A Fable for Tomorrow" is added to one draft in pencil, and the bulk of the language becomes essentially the same as the chapter contemporary readers are familiar with. The process of making the town a composite has led first to the generalization of facts into fiction, then to the expansion of that generalization into a detailed fictionalized framework, and finally to the condensation of the essential elements of that framework into what might be termed a contemporary myth.

Folder 4: Author's Original Typescript of Chapter 1,
Silent Spring (for the New Yorker)

Silent Spring first ran as a series in the *New Yorker* magazine, which was becoming famous for its exposés and articles on environmental topics. In the manuscript prepared for the magazine, Green Meadows retains its name. The *New Yorker* section titled "Reporter at Large" is inserted above the title *Silent Spring,* but the chapter title, "A Fable for Tomorrow," does not yet appear, and neither does it appear in the final *New Yorker* article. Thus, the *New Yorker* version of the chapter provides a concrete example of an option Carson did not choose for the book and reinforces the impression that her choices were deliberate ones.

Folder 5: Typescript of Chapter 1 (with Illustration Marks)

The contents of Folder 5 indicate a growing momentum toward the completion of the book, as finishing details are added. Carson's illustrators, Lois and Louis Darling, worked closely with her to select the style, placement, and frequency of *Silent Spring's* illustrations. Drawn in a simple, dry-brush style, the various landscapes, wildlife, and juxtapositions of technology and nature were scattered throughout the chapter headings, margins, and chapter ends. In this typescript, illustration markers are indicated above the title for chapter 1 ("Landscape—Green Meadows") and at the bottom of that page ("Berries and birds in winter"). A third illustration, an "Empty Feeder," is to be placed next to the lines describing pesticide paralysis. Solely through description and visual illustration, and without having to repeat the phrase "silent spring" once in the body of the text, Carson, her editors, and her illustrators subtly but effectively reinforce the notion of a world without the song of birds.

Folder 6: Revised Draft for the Book

Then the book receives its final editing. The illustrations have been completed, the epigraphs prepared, the typeface selected. A sample page has been run and attached to the front of the manuscript in Folder 6. Page numbers have been assigned and titles of chapters decided upon. Yet it is not until this point that an important—and probably crucial—decision about the fiction in chapter 1 is finally confronted; that is, whether to make any effort to particularize the town by retaining the name "Green Meadows," or finally to raise its status to the level of universal myth.[3]

Carson chooses the latter. Green Meadows becomes not a name but a place, an archetypal American small town. All the references to Green Meadows are subsequently changed to such nouns as "the countryside" or "this town." In the process, a sonorous, repetitive, almost scriptural quality is imparted to the language. The tone resembles that of a parable, or indeed a fable, but a fable with universal—hence, mythical—overtones.

Finally, after all references to the name of the town are excised, Carson insists upon one last major alteration in format: a double space separating the last line of the mythic narrative from the following and final two paragraphs in which Carson contends, in her "own" voice, that the threat is real. With this move, Carson comes full circle, separating fiction and fact once again. But perhaps this is only because the fiction has become a myth, and this myth has expanded beyond any possibility of grounding it in a collection of current events. The myth is spatially liberated to stand on its own; it is both Carson's and the readers' task to note its application to particular daily lives.

I believe it is this striving toward the mythic, toward the most general form of narration, that finally so angered and entranced Carson's readers. A simple story about a particular town, no matter how fictionalized, could be dismissed as irrelevant to anyone's own particular life; a mythic narrative about a town that could be anywhere and a blight that could hurt anyone cannot be ignored. The prospect in such a myth is frightening, awesome, and in its own way, sublime.

Folder 7: Printer's Typescript

Thus far, the development of the manuscript has been so thorough that changes to the printer's typescript are minimal. "Green Meadows" is gone; the double space has been inserted. Carson (most probably) does not like the repetition of "As yet" and "Yet" at the start of two consecutive sentences in the penultimate paragraph, and this is noted. The manuscript goes to press.

Folder 8: Galley Proofs of Silent Spring

Corrections are expensive at this point, but Carson is a meticulous stylist, so they must be made. The offending "As yet" is removed, streamlining the sentence. Much more significantly, Carson makes two important changes: one clarifying and defining the amount of factuality her fiction can bear; the other placing her own personal voice firmly within a nonfiction framework so she can carry on with the rest of her book. Both moves reinforce the original supposition that this fictional narrative is firmly supported by science, but now the appeal of the underlying scientific facts has been enhanced by their association with mythic narrative.

In the first move, Carson must deal with a troubling word, *imaginary*. The problem is that what she describes is imaginary and factual at the same time. But the word *imaginary* connotes untruth to those who do not believe that myth can tell truths. Thus, Carson excises the offending term *imaginary* and substitutes a disclaimer about the town's "actual" existence, as if to clarify the precise nature of its mythical status. Of course the town is imaginary; we know this already, precisely because of its mythic, literary form. But what we might not know, and what Carson strives to maintain, is that the imaginary may also be factually true, despite its lack of "actual" existence. By making this change, Carson informs the reader of the degree to which she is willing to make claims for the factuality of her mythic construction, while she avoids the problems that using the word *imaginary* might evoke.

Later, Carson again replaces the word *imaginary*, this time with the word *imagined*. She does this as if wishing to retain the idea that the imaginary is not necessarily the unreal. Once again, the unfortunate connotation of the word *imaginary* leads Carson to substitute a more neutral, though related, term. One gets the impression that she would still like to use the word *imaginary*, both here and before (as she has through many drafts), but she would rather not be misread at such an important juncture.

In her second move, Carson removes one instance of the first person, further subordinating her viewpoint to that of the problem she is describing. This tragedy is not just in her imagination; it is, instead, a collective nightmare. The movement is toward the general and the mythic. It is significant to note, however, that she does not eliminate herself as narrator completely. Enough of her own viewpoint remains to give the myth a voice, and it is a factual voice, not that of a fictionalized narrator. This is the voice, of course, that must narrate the rest of the book, the nonfictional chapters of *Silent Spring*. Again, because the myth must not be misunderstood as idiosyncratic or untrue, Carson's personal voice is minimized and relegated to the factual domain. Perhaps just such a reformulation of the imaginary, and just such a de-emphasis upon the creative intervention of the author, succeeded in making the mythification of Green Meadows both more acceptable and more disturbing to her readership.

Generalizations and Conclusions

Every rhetorical situation is a particular one. Carson was writing in an era significantly different from our own: the environmental crisis was less perceptually apparent; there was a much greater faith in science as a benign enterprise; and almost no generic precedents existed, except perhaps that of the muckraking journalists of the turn of the century (Eilers; Hynes, chap. 3). Perhaps these situational factors account for the urgency, even the stridency, of her chosen format and style.

But a few generalizations can be derived from this excavation of Carson's process. The first, and perhaps most obvious, is that the relationship between fact and fiction is a complex one and that this complexity, particularly in environmental writings, can be acknowledged and inscribed into the final product. The story of Carson's developmental process underlines her careful weighing and balancing of factual and fictional nuances. Even then, many readers misinterpreted the status of her claims.[4] Her defense was that she had used factual information to build her fictional representation and that every detail in that fiction could be referenced to an episode or event that had already occurred. Moreover, she tried to transmit this relationship of fiction and fact in the very form of her writing. Without overwhelming her story with qualifications and footnotes, she carefully selected words and phrases that spoke to a lifetime of research and experience. She even named the weeds that run by the roadside (laurel, viburnum, and alder), names that the fictional inhabitants of the small rural town might not know. The lesson to be learned here is that the relationship between fact and fiction in any environmental writing is central to its nature and can be made part of the story.

Second, the sequence of Carson's revisions suggest that there is a danger in being too specific, especially in the realm of fiction. Carson's choice not to write a story about a male character touring a town called Green Meadows in 1965 was crucial for reaching a wider audience, both in her own time and today. This is not to say that real and particular locations—whether it be Walden Pond, Sand County, or Arches National Park—cannot be potent sites for focusing environmental concern. It is to say that for environmental writing to touch a wide public, some generalization of particular places is useful and may even be necessary. Further, the kind of generalization one chooses influences the way readers receive the message. Naming a town Green Meadows is an invitation to allegory, tending toward polemic; leaving it unnamed is an invitation to mythology, which inspires contemplation.

Commentators often use the word *allegory* to describe the first chapter of *Silent Spring*. Allegory, however, requires the personification of abstract values or principles (e.g., Good Deeds saves Everyman from the fiery pit; the Slough of Despond keeps Pilgrim from Jerusalem). The extreme result can be a leaching out of any realistic detail or characterization, with the story line seeming to function only on the abstract level. In its final version, "A Fable for Tomorrow" does not go so far—the town is not an abstraction (Green Meadows) but a composite of factual events happening in existing towns, as the text makes clear. Moreover, real birds and piglets are dying (not the Spirit of Spring or Fecundity).

Alternatively, the realistic end of the literary spectrum is the novel. The events and persons in a novel are "really" doing things, and their greater significance, while important, is subordinated. The manuscript drafts of "A Fable for Tomorrow" indicate a novelizing tendency in the character of the young man returning home. This touch added concrete detail but not much significance. So by excising the character, Carson moved toward the generalizing end of the scale but not so much as to lose the immediacy and urgency of the verifiable events upon which the story was based.

Myth may be both a better descriptor than *allegory* for the generalized elements in the chapter and a better strategy than *novel* for retaining the realist elements. Myths retain their concrete, material quality (e.g., Cronos castrates Sky) while at the same time they can be abstracted (e.g., Time conquers boundless Space). In the case of *Silent Spring*, the town retains its quality as a material entity even while standing for all towns that could be poisoned by pesticides. The choice of a given genre may depend as well upon particular readers' readiness to receive an environmental message. In this case, it may have been too early to expect readers to identify with a particular location, even a fictional one.

The third and final generalization this essay proposes is that the voice of the author plays a key role in identifying the status of the fictionalized content. Often the presence of a first-person narrator disqualifies the content of writing as being truly "factual" or "objective," as in most scientific genres. But often the lack of an explicit authorial presence hides or subsumes personal responsibility for the facts and how they are manipulated rhetorically. Carson's solution was to objectify her fiction and personalize her facts. That is, by removing a first-person homodiegetic narrator from her fiction, Carson underscored the universal truths of the myth she had created; and by associating a heterodiegetic voice with the nonfictional elements of her message, she took responsibility for its factuality. Environmental writing, which often draws upon the writer's experience as a source of information and judgment, must often strike such a balance between the "objective" and "subjective" modes without overwhelming one or the other.

As in every rhetorical situation, Carson's choices were constrained by generic, audience, and stylistic considerations. Whether these choices were the right ones—we know they were successful ones—is at this stage in history beside the point. In any case, an examination of the invention of "A Fable for Tomorrow" outlines a remarkable story of the inventional processes involved in one of the most successful rhetorical accomplishments of our century and perhaps the most significant one for environmental activism. The process of excavation, for whatever purpose, can only help us understand the working habits of an expert, a craftsperson well worth imitating. And perhaps understanding how these choices were made can be of some help when, as students of communication and advocates for the environment, we confront the myriad inventional problems endemic to our own place and time.

Notes

I wish to thank the National Endowment for the Humanities and the University of Utah College of Humanities Development Committee for the grants that enabled this project. I also wish to thank Carson biographer Linda Lear, research fellow at the Smithsonian Institution Archives, for her invaluable contributions.

1. This interpretive process suggested itself when I called the folders, in sequence, from the well-organized Carson collection and was surprised and intrigued by each new discovery.

2. I wish to thank Linda Lear for calling my attention to this article, of which Carson was aware (e-mail, 14 May 1996).

3. Paul Brooks, for one, was against the idea of retaining the name of the town. He felt that "Green Meadows" "suggests not an old town but a real estate development" (Graham 64).

4. Linda Lear, in a personal communication to me (e-mail, 20 Feb. 1996), wrote that many of Carson's friends affirmed that she was frustrated by the

controversy generated by "A Fable for Tomorrow" because it drew attention away from her main intent.

Works Cited

Booth, Wayne C. *The Rhetoric of Fiction.* 2nd ed. Chicago: U of Chicago P, 1983.

Brooks, Paul. *The House of Life: Rachel Carson at Work.* Boston: Houghton, 1972.

Carson, Rachel. Papers. Yale Collection of American Literature MSS 46, Box 38, Folder 680 ("Folder 1"); Box 50, Folder 899 ("Folder 2"); Box 53, Folder 900 ("Folder 3"); Box 53, Folder 967 ("Folder 4"); Box 53, Folder 975 ("Folder 5"); Box 54, Folder 987 ("Folder 6"); Box 55, Folder 1005 ("Folder 7"); Box 55, Folder 1014 ("Folder 8"). Beinecke Rare Book and Manuscript Library, Yale University.

———. *Silent Spring.* 1962. 25th anniv. ed. Foreword by Paul Brooks. Boston: Houghton, 1987.

———. "To Dorothy Freeman." 13 June 1961. *Always, Rachel: The Letters of Rachel Carson and Dorothy Freeman, 1952–1964.* Ed. Mary Freeman. Boston: Beacon, 1995. 380–81.

Eilers, Perenthia. "Creating an Environmental Conscience: Revelation and Depiction as Rhetorical Strategies in Rachel Carson's *Silent Spring.*" Master's thesis. U of Wisconsin, 1994.

Foucault, Michel. *The Archaeology of Knowledge and the Discourse on Language.* Trans. A. M. Sheridan Smith. New York: Pantheon, 1994.

Graham, Frank, Jr. *Since Silent Spring.* Boston: Houghton, 1970.

Hynes, H. Patricia. *The Recurring Silent Spring.* New York: Pergamon, 1989.

Killingsworth, M. Jimmie, and Jacqueline S. Palmer. "The Discourse of 'Environmentalist Hysteria.'" *Quarterly Journal of Speech* 81 (1995): 1–19.

Lear, Linda. E-mail to the author. 14 May 1996.

———. E-mail to the author. 20 Feb. 1996.

———. *Rachel Carson: Witness for Nature.* New York: Holt, 1997.

Terres, John K. "Dynamite in DDT." *New Republic* 25 Mar. 1946: 415–16.

4

EDWARD P. J. CORBETT

A Topical Analysis of "The Obligation to Endure"

৯৯ In 1962, long before the pollution of our environment became a prominent public issue, Rachel Carson published *Silent Spring,* an early warning about the dangerous contamination of the environment by the indiscriminate use of chemical insecticides. To assess her effectiveness as a persuader, one would have to examine the entire book, because it is the book as a whole that presents her case fully. Here, however, I examine only an excerpt from the book, chapter 2, "The Obligation to Endure," noting Carson's rhetorical strategies in general and speculating about the topics that have yielded her arguments.

The topics, or *topoi,* are the general heads under which classical

An earlier version of this essay appeared in *Classical Rhetoric for the Modern Student,* fourth edition, by Edward P. J. Corbett and Robert Connors. Copyright © 1965, 1971, 1990, 1999 by Oxford University Press, Inc. Used by permission of the author and Oxford University Press.

rhetoricians—such as Aristotle, Cicero, and Quintilian—grouped arguments for a particular subject or occasion. They are the regions, the haunts, the places where certain categories of arguments reside. Collectively, they function as a checklist of ideas on a given subject. Initially, writers or speakers might review the entire list of topics, asking whether each one suggests material for the development of their subject. Eventually, however, they may discover that many topics are unsuitable for certain situations. As Quintilian said,

> I would also have students of oratory consider that all forms of argument which I have just set forth [i.e., the topics] cannot be found in every case and that when the subject on which we have to speak has been propounded, it is no use considering each separate type of argument and knocking at the door of each with a view to discovering whether they may chance to serve to prove our point, except while we are in the position of mere learners. (5.10.122)

Quintilian envisioned the time when, as the result of study and practice, students would acquire a "power of rapid divination" that would lead them directly to those arguments suited to their purposes. In addition to serving as a heuristic for the *construction* of rhetoric, the topics can also serve as a heuristic for the *analysis* of rhetoric, providing insight into the effectiveness of such discourse.

Classical rhetoricians commonly divided the topics into two general kinds: common topics and special topics. Common topics are depositories of general arguments that one can resort to when discussing virtually any subject. The common topics and their subtopics are as follows: (1) Definition: Genus and Division; (2) Comparison: Similarity, Difference, and Degree; (3) Relationship: Cause and Effect, Antecedent and Consequence, Contraries, and Contradictions; (4) Circumstance: Possible and Impossible and Past Fact and Future Fact; and (5) Testimony: Authority, Testimonial, Statistics, Maxims, Law, and Precedents (Examples).

Special topics, on the other hand, are more particular lines of argument that one can resort to when discussing some particular subject. In the three main kinds of rhetorical activity—deliberative, judicial, and ceremonial—we tend to rely, fairly regularly, on special topics, which are more particular than common topics and yet more general than specialized knowledge. In a case at law, for instance, the teacher of rhetoric could not be expected to instruct students in all the points of law peculiar to this case; but the teacher could point out to students that most law cases turned on a limited number of recurring topics.

"The Obligation to Endure," chapter 2 of Rachel Carson's *Silent Spring,* is clearly an example of deliberative discourse, an attempt to change the attitudes and actions of the audience in regard to a matter of public concern. Like all deliberative discourse, this one deals, ultimately, with the future. Of course, it examines the contemporary situation, but it does so for the purpose of effecting change in public policy in the future—the *near* future. Hence, I turn now to the special topics for deliberative discourse.

There are some common denominators among the appeals that we use when we are engaged in exhorting someone to do or not to do something, to accept or to reject a particular view of things. When we try to persuade people to do something, we try to show them that what we want them to do is either good or advantageous. All of our appeals in this kind of discourse can be reduced to these two heads: (1) the Worthy (*dignitas*) or the Good (*bonum*) and (2) the Advantageous or Expedient or Useful (*utilitas*). The English terms do not express fully and precisely what the Latin terms denote, but perhaps they do convey the general sense of "what is good in itself" (and therefore worthy of being pursued for its own sake) and "what is good for us" (a relative good, one that would be expedient for us to pursue because of what it can do for us or what we can do with it). John Henry Newman used somewhat the same distinctions in his *Idea of a University* when he was trying to establish the difference between "liberal knowledge" and "useful knowledge." There are some subjects or disciplines, Newman maintained, that we cultivate for their own sakes, irrespective of the power they give us or the use to which they can be put. There are other subjects or disciplines that we cultivate primarily, if not exclusively, for the use to which we can put them.

If we were trying to convince someone to study poetry, for instance, we might urge that the cultivation of poetry is a good in and of itself and therefore worthy of pursuit for its own sake and that it is no depreciation of the worth of poetry to admit that the study of poetry cannot be put to any practical uses. On the other hand, we might conduct our appeal on a less-exalted level by showing that the study of poetry can produce practical results. The study of poetry can teach us, for instance, how to be more intensive readers, how to be more precise writers, and how to be keener observers of the world about us.

Whether we lean heaviest on the topic of the Worthy or the topic of the Advantageous will depend largely on two considerations: the nature of our subject and the nature of our audience. For instance, it is easier to demonstrate that the study of poetry is a good in itself than to demonstrate that the building of a bridge is a good in itself. In the latter case, it would be a wiser course—because it is the easier and more cogent course—to exploit the topic of the Advantageous. When one is trying to convince a group of taxpayers that they should vote for a bond issue to finance the building of a bridge, one is more likely to impress the taxpayers by demonstrating the usefulness of such a bridge than by showing its aesthetic value.

So too the nature of our audience will influence our decision about the special topic to emphasize in a particular deliberative discourse. Because the audience is the chief determinant of the best means to effect a given end, we should have at least a general sense of the temper, interests, mores, and educational level of our audience. If we discern that our audience will be more impressed by appeals based on the topic of the Worthy, then our knowledge of the audience will have to be a little fuller and a little more accurate because now we must have, in addition to a general sense of the temper of the audience, some knowledge of just what things are regarded as "good" by this audience and what the hierarchy of good things is. People being what they are, however, it is probably true, as Cicero (*De Inventione; Topica*) and the unknown author of the *Rhetorica ad herennium* observed, that the topic of advantage will appeal more frequently to more people than will the topic of worth.

All deliberative discourses are concerned with what we should choose or what we should avoid. Aristotle observed that the end that determines what people choose and what they avoid is happiness and its constituents. Happiness then may be looked upon as the ultimate special topic in deliberative discourse since we seek the worthy and the advantageous because they are conducive to our happiness. Unquestionably, people differ in their concept of happiness. However, Aristotle presented some of the commonly accepted definitions:

> We may define happiness as prosperity combined with virtue; or as independence of life; or as the secure enjoyment of the maximum of pleasure; or as a good condition of property and body, together with the power of guarding one's property and body and making use of them. That happiness is one or more of these things, pretty well everybody agrees. (1.5)

Even today, most people, if asked, would say that happiness was one or another of these conditions or perhaps some combination or modification of these. They would probably agree too with Aristotle's specifications of the constituent parts of happiness: "good birth, plenty of friends, good friends, wealth, good children, plenty of children, a happy old age, also such bodily excellences as health, beauty, strength, large stature, athletic powers, together with fame, honor, good luck, and virtue."

In summary then, when we are engaged in deliberative discourse, we are seeking to convince someone to adopt a certain course of action because it is conducive to happiness or to reject a certain course of action because it will lead to unhappiness. The two special topics under the general head of happiness are the Worthy and the Advantageous. In developing these special topics, we will sometimes have occasion to use some of the common topics, such as the Possible and the Impossible (when urging the advantage, for instance, of a certain course of action, we may have to show that the course we are advocating is practicable or easy) and the topic of more or less (when seeking to direct a choice from among a number of goods, for instance, we need criteria to help us discriminate degrees of good). In developing the special topics, we will also need a fund of specialized knowledge pertinent to the subject we are debating. In discussing matters of public polity, for instance, we need a great deal of accurate and specific knowledge about kinds of government, the laws and constitution of a state, the mechanism of legislation. In addition, we must have at our fingertips, or know where to find, a sufficient number of facts, precedents, or statistics to support our assertions. I turn now to a topical analysis of "The Obligation to Endure."

Of the special topics that prevail in deliberative discourse—the Good and the Unworthy, the Advantageous and the Disadvantageous—Rachel Carson relies heavily on the Advantageous and Disadvantageous. To swing us away from our present practices, she has to persuade us that those practices are disadvantageous, that they are detrimental to the well-being of society, and that the practices she advocates are advantageous, that they will mitigate or eliminate the detrimental effects of our present policy. Before we examine how Carson argues in this piece of deliberative discourse, we should get an overview of the main structure of this second chapter of her book:

Introduction (paragraph 1)
I. Exposition of the situation in which people have developed an enormous capacity for contaminating the environment through nuclear radiation and chemical sprays (paragraphs 2–12)
II. Exposition of the situation in which the remedies invented to counteract disease-carrying and crop-destroying insects became worse than the problem (paragraphs 13–25)
Conclusion (paragraph 26)

As this overview makes clear, Carson has devoted just about the same number of paragraphs to each of the two main, roman-numeral parts of the discourse. The first part defines what the problem is; the second part explains how the problem developed. We might say that there is a cause-and-effect relationship between the two parts—part one being the effect, part two being the cause. And we will see later that the topics of Cause and Effect, Antecedent and Consequence have yielded many of Carson's arguments.

In the very first sentence of paragraph 1, which constitutes the introduction to this excerpt, Carson enunciates the premise upon which she bases her exposition of the development of the problem: "The history of life on earth has been a history of interaction between living things and their surroundings" (5). Then she quickly points out a shift that has taken place in this basic relationship. Whereas for centuries the environment held predominant power over living things, in the present century, one species of living things, human beings, has assumed predominant power over the environment.

In the next paragraph, which begins the body of her argument, Carson argues that the Difference in the relationship between environment and living things is a difference not only in degree but in kind. Not only has the power of human beings over the environment increased at an accelerating rate, but that power is now exerting a lethal, irreversible influence on the environment. Mainly through the release of nuclear radiation and chemical sprays, people have pervasively contaminated the environment in which they live. At the end of this paragraph, Carson quotes Albert Schweitzer, providing an argument by Maxim: "Man can hardly even recognize the devils of his own creation" (6).

In paragraph 3, Carson points out another Difference in the situation that once prevailed and the one that now prevails. In the past, there was ample time for life to adjust to the natural changes in the environment; but today, the time needed for that kind of adjustment is not available. By contrasting the past with the present, Carson might be said to be exploiting the topics of the Possible and the Impossible for the future.

In the following sequence of paragraphs (4–8), Carson explores the reasons why we no longer have the necessary time for adjustments (and here the arguments are yielded mainly by the topic of Cause and Effect or the topic of Antecedent and Consequence): (1) the acceleration of the pace of contamination (paragraph 4); (2) the acceleration of the rate at which new chemicals are created (paragraphs 5–6); (3) the acceleration of the rate of indiscriminate destructiveness by insecticides (paragraph 7); and (4) the acceleration of the degree of toxicity in the new sprays (paragraph 8).

In paragraph 9, Carson provides a Definition of "the central problem of our age" (that is, a problem over and above the ever-present danger of the extinction of humanity by nuclear warfare): how to prevent the contamination of the whole environment with chemical sprays (8). An argument from Future Fact is introduced rather incidentally in paragraph 10: not only can chemical sprays pollute our environment, but they also have the potential, like nuclear radiation, of causing gene mutations.

In paragraph 11, Carson raises the provocative question that was mainly responsible for her writing not only this chapter but the whole book: Why, in seeking to control a few unwanted species, have we risked the contamination of our entire environment? She then refutes *one* of the arguments commonly offered for the use of chemical sprays—namely, the argument that we must use insecticides to maintain our farm production. She exposes the weakness of that argument by pointing out that our present farm production is so high that we have to spend billions of dollars every year to store the excess.

In paragraph 12, the final paragraph in the first main division of the discourse, Carson enunciates her thesis: not that we do not need to control the insect population but that the "control must be geared to realities, not to mythical situations, and that the methods employed must be such that they do not destroy us along with the insects" (9). That is the policy she will be advocating farther on in the book, after she has convinced us of the clear and present danger of the prevailing situation. In the next major division of the essay (paragraphs 13–25), she will go on to point out for us what the "realities" are so that she can counteract the "mythical situations" that have dictated our practices so far.

In this first major division, Carson's rhetorical strategy has been to start out with a rather general proposition and then gradually narrow down the discussion to the central problem that she is primarily concerned with. She starts out with the general premise that there is a natural interrelationship between living things and their environment. Then she points out that there has been a profound shift in the predominance of the two factors in this relationship. Then she focuses on just two of the ways in which people have assumed predominance over the environment—through nuclear radiation and through chemical sprays. By the time she reaches the end of this first major division, she has narrowed down the discussion to just one form of chemical sprays—insecticides. And for the remainder of this chapter, she will concentrate on insecticides.

In the second main division of the discourse, Carson is engaged primarily in presenting the historical and biological facts that account for the insect population becoming a threat to humanity. She will argue, however, that the method of control that people devised—spraying chemical insecticides into the atmosphere—has worsened the condition it was intended to correct. The topics that yield most of her arguments in this second section are Cause and Effect, Antecedent and Consequence, Testimony, and Statistics.

In paragraph 13, after making the point that insects inhabited the earth long before people did, Carson resorts to the topic of Division to make the point that eventually a small percentage of insects became a threat to humanity in two principal ways: as competitors for the food supply and as carriers of disease. In paragraph 14, she deals briefly with one of these threats: disease-carrying insects. After picturing the situations that intensify the threat from disease-carrying insects, she concedes that some control of these insects is necessary, but she contends that the remedy people devised is worse than the disease. There is an element of the topic of Contraries in her argument here, an argument that runs something like the following: disease is *bad*; a remedy is supposed to be *good*; but the remedy in this case is worse than the disease.

From here to the end of the discourse, Carson deals with the second of the threats: those insects that compete with human beings for the food supply. In paragraph 15, she resorts to Comparison to point out the Difference between the primitive agricultural situation and the agricultural situation today. Unlike the primitive agricultural situation, which had its own built-in system of checks and balances, single-crop farming today has intensified the growth of the insect population. In the next paragraph (16), she resorts to

Comparison again to point out the analogous situation in which there was a wholesale planting of a single species of trees (elms) in many cities of the United States at the beginning of the twentieth century.

In paragraphs 17–19, Carson deals with another factor that has contributed to the intensification of the growth of the insect population in our time: the emigration of thousands of new species of insects from their native habitat to the United States. She cites the Authority of the British ecologist Charles Elton for an account of how this influx took place in the natural course of events as a result of the geological separation and then rejoining of masses of land by the action of the seas. In paragraph 18, she relates how this influx took place accidentally as the result of insects hitchhiking on plants imported into this country. She cites the Statistic that almost half of the 180 major insect enemies of plants were introduced into the United States in this way. In paragraph 19, she resorts to an argument from Degree, saying that invading insects are often more devastating than native varieties because in their new environment, free of the natural checks and balances of their native habitat, they proliferate enormously.

In paragraph 20, she presents an argument from Antecedent and Consequence. Since these invasions are likely to continue indefinitely, we need to know more about animal populations and their relations to their environment. She quotes a Testimonial from Charles Elton's book again about this matter. In the next paragraph (21), she mentions the ironic fact that, although much of this necessary knowledge is now available from entomologists and ecologists, most of us do not heed this Authority. She raises the question, in paragraph 22, of why we have accepted as inevitable "that which is inferior or detrimental, as though having lost the will or the vision to demand that which is good" (12). She quotes a Testimonial from the ecologist Paul Shepard about our inexplicable acceptance of an intolerable ecological situation.

In the next paragraph (23), Carson resorts again to an Antecedent–Consequence argument, maintaining that we have been so brainwashed about the need for a chemically sterile, insect-free world that we have enfranchised control agencies to exercise their powers ruthlessly and indiscriminately. She quotes the Authority of Neely Turner, a Connecticut entomologist, to the effect that the most flagrant abuses of the "regulatory entomologists" have gone unchecked by state and federal agencies (12).

In paragraphs 24–25, the final two paragraphs of the second major division, Carson repeats the thesis of this chapter and the whole book: "[W]e have put poisonous and biologically potent chemicals indiscriminately into the hands of persons largely or wholly ignorant of their potentials for harm" (12). In stating her thesis for the second time, she wants to make clear to her readers what she is advocating: not a categorical rejection of the use of chemical insecticides but a responsible and discriminating use of these lethal chemicals.

In the concluding paragraph (26), Carson winds down this section of her argument. What she mainly wants to do in this paragraph is suggest a policy that the public should adopt in response to the menace of polluting insecticides: "The public must decide whether it wishes to continue on the present road, and it can do so only when in full possession of the facts" (13). The quotation (almost a Maxim) from Jean Rostand with which she concludes the paragraph—"The obligation to endure gives us the right to know"—not only provides her with the title of this chapter but suggests that the decision that she is asking the public to make is not a matter of indifference but a matter of obligation. A Consequence of our obligation to endure is a right to know the facts. And in the rest of the book, she will be supplying her readers with the specific facts about chemical insecticides that will help the public make the obligated decision.

Throughout this chapter, Carson has raised the apprehensions of her readers about the potential dangers of an indiscriminate and irresponsible use of chemical sprays. She has prepared them to receive the damaging facts that she will present in the rest of the book through Testimony of experts, through Examples and Statistics. As we know from subsequent history, Rachel Carson succeeded in raising the consciousness of a large segment of the American public about the dangers of fouling our own nest: her arguments have been effective.

Works Cited

Aristotle. *Art of Rhetoric.* Trans. J. H. Freese. Cambridge: Harvard UP, 1959.

Carson, Rachel. *Silent Spring.* Boston: Houghton, 1962.

Cicero. *De Inventione.* Trans. H. M. Hubbell. Cambridge: Harvard UP, 1949.

———. *Topica.* Trans. H. M. Hubbell. Cambridge: Harvard UP, 1949.

[Cicero.] *Rhetorica ad herennium.* Trans. Harry Chaplan. Cambridge: Harvard UP, 1954.

Newman, John Henry. *The Idea of a University.* 1852. Notre Dame: U of Notre Dame P, 1982.

Quintilian. *Institutio Oratoria.* Trans. H. E. Butler. Vol. 2. Cambridge: Harvard UP, 1920–22.

5

TARLA RAI PETERSON AND
MARKUS J. PETERSON

Ecology According to *Silent Spring's* Vision of Progress

&❧ Rachel Carson's *Silent Spring* played a significant role in transforming environmentalism into a central aspect of the public consciousness. Carson's radical arguments—that care for the environment is intrinsically good, that nature provides the basis for all technological innovation, and that the quality of human life depends on sustaining that foundation—generated conflicts that rage to this day. In this essay, we explore connections between *Silent Spring* and contemporary ecology and environmental-science textbooks, arguing that much of *Silent Spring's* vocabulary has been incorporated into mainstream environmental-science education. More important, Carson's claim that the very existence of a warrior mentality (against insects or other pests) is unhealthy for humans and the planet seems to have been accepted by the authors of these texts.

An earlier version of this essay appeared in the 1997 *Proceedings of the Conference on Communication and Environment.* Edited by Susan L. Senecah. Used by permission of the author.

Silent Spring offers a metaphorical system that encourages readers to understand the world, including humans, as a vast tapestry of interconnected threads. Much has been written about the apocalyptic tone, activist nature, and emotive power of its discourse. Opie and Elliot write of the evocative language and apocalyptic vision in "A Fable for Tomorrow." Killingsworth and Palmer characterize *Silent Spring* as "consciousness-raising" ("Liberal" 220–21) and a "visionary polemic" ("Millennial Ecology" 22), while Oravec argues that it defines Carson's environmental activism.

Carson also offers traditional scientific evidence for her claim of interconnectedness. She integrates "scientific findings with anxieties over the earth's capacity to support the growth of a technologically advanced civilization" (Killingsworth and Palmer, "Millennial Ecology" 27). *Silent Spring* suggests that, for reasons deriving wholly from *logos,* chemically driven attempts to manage insect pests are inappropriate. Carson's use of logical appeals made it easy for environmental-science and ecology texts to incorporate many of her arguments, as well as the metaphors that give them a sense of urgency. By simultaneously appealing to the intellect and the emotions, Carson lent the authority of science to those who questioned the chemical industry's expanding influence in the management of nature.

Our essay is grounded in Kenneth Burke's notion of rhetorical selectivity *(Language, Rhetoric),* and bounded by Niklas Luhmann's theory of function systems *(Ecological Communication,* "What is Communication?"). Burke claims that the terms, or vocabulary, that make up our rhetoric serve as "terministic screens," emphasizing some aspects of reality while de-emphasizing others. According to Luhmann, contemporary society is best conceptualized as a loosely coupled set of function systems that recognize each other's existence only through reliance on intra- and intersystem communication. Together, these approaches guide our analysis.

Rhetorical Selectivity and Constraints

Burke asserts that the chief end of rhetoric is *identification*, or a sense of unity with our fellow humans *(Rhetoric)*. Burke's theory of rhetoric focuses on communication as the process whereby we humans overcome our alienation from each other. We use rhetoric as a symbolic means for expressing the common interests needed to achieve the social unity we desire (Burke, *Language, Rhetoric*). *Silent Spring* functions rhetorically as it offers symbols that encourage otherwise diverse individuals to achieve unity. It encourages readers who share a desire to prevent the devastation threatened in Carson's "A Fable for Tomorrow" to discover ways they can work together to sustain the natural environment.

Burke argues that critical discourse provides a means for exposing the assumptions that have become implicit in our cultural stories *(Attitudes)*. The critic can take either a tragic or a comic approach to this task. For example, Carson's fable of mass destruction offers a tragic interpretation of the quest to improve humanity's quality of life and standard of living wherein the chemical crusaders are destroyed by their own hubris. Conversely, Carson's call to action suggests a more comical assessment of humanity. She asks, "[H]ow could intelligent beings seek to control a few unwanted species by a method that contaminated the entire environment and brought the threat of disease and death even to their own kind?" (8). Quoting the ecologist Paul Shepard, she points out the silliness of choosing to "live in a world which is just not quite fatal," and she urges her readers to arise from their "mesmerized state" and make decisions "in full possession of the facts" (12–13).

Burke suggests four interrelated concepts that guide our exploration of *Silent Spring*: an *occupational psychosis*, which is constituted through a *terministic screen*, results in *trained incapacity*. *Perspective by incongruity* then clarifies the role of criticism in fostering appropriate revisions to the social hierarchy. An occupational psychosis is a way of thinking that develops out of a certain pattern of living, then reinforces continued reliance on traditional patterns in nontraditional conditions (*Permanence* 37–50). Burke recognizes that, while no society is homogeneous in its occupational psychosis, certain occupations are more characteristic of certain societies than are others. He argues that contemporary Western society suffers from a "technological psychosis," which he describes as an expectation that technological progress can transcend human mortality by achieving absolute control over nature. Although Burke endorses neither the desirability nor the practicality of this goal, he claims it provides the patterns we use to measure the desirability of all social activity.

Individuals and the societies within which they live develop a vocabulary, or a *terministic screen,* to make sense of their occupational psychoses. Over time, people weave individual terms together, forming a screen that enables its users to decide which aspects of experience are important, what that experience means, and what sort of action it calls for. Thus, a terministic screen provides a meaning system that constrains our ability to turn reality into information, provides tools for evaluating and naming situations, and encourages us to adopt certain roles within those situations. As an occupational psychosis diffuses throughout a culture, society selects certain vocations as more central than others. This choice is reflected and furthered by the dominant terministic screens. Occupational psychosis combined with nonreflective use of its accompanying terministic screen leads to *trained incapacity,* or a condition in which our abilities "function as blindness" (*Permanence* 7–11, 49). Because our terministic screens constrain our observational possibilities to those in keeping with our occupational psychoses, we see new experiences in the terms provided by our past training. If conditions have radically changed since our terministic screen was developed, our training becomes an incapacity. In this sense, our occupational psychosis works as a de-skilling process that disables us from seeing new alternatives for dealing with situations.

Neither society nor individuals could function without terministic screens. Burke points out that "to explain one's conduct by the vocabulary of motives current among one's group is about as self-deceptive as giving the area of a field in the accepted terms of measurement. One is simply interpreting with the only vocabulary he knows" (*Permanence* 21). Difficulties arise because, rather than reflect on their terministic screens when faced with a new problem, people interpret "*the problem in such a way that [their] particular aptitude becomes the 'solution' for it*" (*Permanence* 242–43, his emphasis). One strategy for expanding our capacities is to adopt *perspective by incongruity*. Constructing new and incongruous names for our material circumstances enables us to "remoralize" situations that have been "demoralized" by our trained incapacities. Stylistic revision of our original perspective is insufficient to achieve incongruity; our fundamental point of view must be changed. *Silent Spring* offers an incongruous perspective that invites its readers to re-vision their relationship with the rest of nature.

Luhmann's social theory helps explain how terministic screens simultaneously grow out of and tend to stabilize a society (*Ecological Communication*, "What Is Communication?"). Luhmann proposes that functional relations exist between a problem and a *range* of possible responses; and problems that do not acquiesce to such a range are not social problems. Luhmann's insistence on the continual presence of possibilities complements Burke's argument that the human condition is one of limited choice (Peterson 36–45). Luhmann describes social relations with the environment as internally driven *responses to*, rather than *interactions with*, the environment (*Ecological Communication* 989). In other words, the environmental disaster detailed in "A Fable for Tomorrow" was not the result of interactions between society and its environment so much as it was the result of *internally* driven *social* responses to environmental conditions. Society's rush to control nature, its headlong pursuit of technologically defined progress, and the public's naive trust in technological interpretations of the natural world combined to produce the disappearance of songbirds, withered vegetation, and puzzling illnesses that gripped farm families. The desolate community Carson describes is the culmination of elements internal to the social world that constrained both awareness of and responses to environmental conditions. The terministic screen through which a society articulates its attitudes toward the environment is more a product of the society's occupational psychosis than a response to environmental conditions.

Luhmann defines society as an "all encompassing social system of mutually referring communications . . . [that] originates through communicative acts alone and differentiates itself from an environment of other kinds of systems through the continual reproduction of communication by communication" (*Ecological Communication* 7). Societal stochasticity, however, maintains possibilities for emergence and evolution of new system identities. The society wherein these communicative transactions occur is a group of function systems that constrain both what can be communicated and how it is communicated. Because modern society recognizes no single authority figure that can cut across all questions and social issues, individual functions assume primary authority for resolving problems. Luhmann identifies economy, law, science, politics, religion, and education as the most important function systems in contemporary society (*Ecological Communication*). These systems constrain both what experience becomes information and what kind of information it becomes. They provide the frames for terministic screens by sorting all experience. However, function systems can utilize terms from within other function systems without losing their identities (Peterson and Peterson, "Valuation"). Even in cases where function systems do not produce coordinated responses, their interdependence ensures that operations can switch quickly from the terms of one function system to the terms of another.

Luhmann cautions against defining the world solely in terms of a single function system (*Ecological Communication*). Terministic screens that privilege one mode of experience over all others threaten to distort the social experience (see Peterson 32–53). For example, when observations of the natural world are interpreted according to the economic function system, the ability and/or inability to pay the costs for preserving the landscape becomes the only decision rule. Only through recognizing the limitations of each function system can society benefit from the internal complexity of the integrated system into which it has evolved.

Silent Spring's Vision of Progress

A New Terministic Screen

Silent Spring offers an alternative to technological psychosis yet does not require readers to reject science. Carson uses scientifically validated information to weave humanity into the vast tapestry of life on earth. She does not insist that all life is the same but leaves considerable room within which a reader can negotiate a place for human needs and desires that is just a bit more special than that occupied by other life. She does not urge people to return to a previous age of innocence but to move forward out of the "stone age of science" (297). She offers a revised view of progress that accounts for multiple perspectives, including, but not limited to, technological solutions to environmental dilemmas. She does not just give us good and evil. Instead, she weaves a terministic screen that accounts for a complex interconnection between humans and other earth life.

Carson characterizes the pro-chemical view as a simplistic "war against nature" (7). In chemicals, she writes, humans possess a "battery of poisons of truly extraordinary power," a battery that has been hailed as the means for "winning the farmers' war" (20). Even if our chemical weapons do not turn on us, we fight at our own peril. Carson writes that insects eventually evolve unnatural strength to match the unnatural weapons thrown against them. After enduring years of spraying, these "unwanted species" have "evolved super races" in "triumphant vindication" of natural over artificial power (8). She uses both the results of traditional scientific studies and detailed stories to show that increasing numbers of insects now "defy our efforts to control them," having become "impervious to spraying" (263–64). The chemical warfare introduced by humans has produced a "true Age of Resistance" because people have failed to realize that insects "possess an effective counterweapon" (263–64). Whereas insects become evermore powerful as we fight them, humans have ceded their power to chemical weapons. Chemicals take a "staggering" toll as they "shatter" natural patterns, leaving a "train of disaster in [their] wake" (7–9). Not only are chemicals powerful, but they are clever. Their residue "lurks in our environment," and "casts a shadow that is no less ominous because it is formless and obscure" (187–88).

Carson is able to offer a surprisingly optimistic vision in the face of such destruction because she portrays most human supporters of chemical warfare as ignorant rather than inherently evil. She writes that the unsuspecting public has been subjected to "tranquilizing pills of half truth" with a "sugar coating of unpalatable facts" (13). Most of the time, they are unaware of reality as they "walk unseeing through the world" (249). Ignorance can be remediated with knowledge, although such a process is rarely easy or pleasant.

There are, of course, people who have an interest in maintaining public ignorance. The technocrats dressed in scientific clothing are Carson's real villains. They have been corrupted by the "giddy sense of power" their "bright new toy" imparts (68). Carson writes that the best of them "operate in the belief that salvation lies at the end of a spray nozzle" (259). When faced with increasing rates of insect resistance, they "blithely put their faith in the development of new and ever more toxic chemicals" (264–65). In their pursuit of power, they subject all life, including humanity, to "senseless and frightening risks" (278). They are clever but not wise.

Silent Spring challenges the reader to abandon the "deceptively easy" road of chemical dependence, for a road "less traveled by" (277–78). Carson does not specify which route we should select, only that "we should look about and see what other course is open to us" (278). She suggests that scientists, rather than technicians, can provide an appropriate model. She characterizes science as emulating nature in its patient deliberateness, its long-term focus, and its life-giving goals. Scientists do not abdicate the human role as director or manager of natural systems, however. Rather, their decisions are "based on understanding of the living organisms they seek to control, and of the whole fabric of life" (278). They seek "new, imaginative, and creative approaches" to "sharing our earth" with other life forces, "cautiously seeking to guide them into channels favorable to ourselves" (296).

Silent Spring creates an alternative vision of progress by wrenching science and technology apart, then providing a new image of both terms. Technology, Carson tells us, is the tool of a rash cadre of warriors who rush imprudently into battle without the slightest understanding of their enemy. Science, on the other hand, is carried out with patient care. Scientists seek to understand nature, including insects, and to learn from it. They seek to help the rest of humanity work out a comfortable position within the natural world. They do not expect humans to endure unnecessary suffering for the benefit of nonhuman life, but neither do they seek to inflict unnecessary damage on that life. These new images for science and technology provide the vocabulary out of which Carson's reader can weave a new terministic screen.

A Multifunctional Perspective

Silent Spring's redefinition of progress has structural, as well as symbolic, implications. Despite its broader ideological roots (Killingsworth and Palmer, "Millennial Ecology"), progress has become increasingly defined in economic terms. Rather than ask her readers to give up progress, Carson redefined the term to include multiple function systems. The fact that Carson redefined rather than rejected progress did not make her work any less controversial. In fact, Carson's claim that decisions about how to manage insect pests had implications that transcended society's economic dimensions may have made her work even more threatening.

Carson's new vision of progress affords importance to multiple function systems, with centrality afforded to science. She used biological science reports, as well as stories from communities all across the United States and other parts of the world, to demonstrate that humans are part of nature and cannot be isolated from the toxic chemicals they pour into the environment. Carson made extensive use of research published in premier science journals, such as *Nature* and *Science,* as well as prominent ecology journals, including *Ecology* and the *Journal of Wildlife Management.* She also relied on leading ecologists and naturalists, such as the distinguished British ecologist Charles S. Elton and the influential U.S. wildlife biologists Olaus J. Murie, Joseph J. Hickey, and Clarence Cottam. Although Carson grounded her vision of progress in science, she carefully wove into the tapestry of her narrative the emotive power of stories about how chemical pesticides affected ordinary people. These stories appear in several places in nearly every chapter. For example, to illustrate how endrin could lead to human poisoning, even when used with apparently adequate precautions, Carson told the story of a U.S. family who moved to Venezuela:

> There were cockroaches in the house to which they moved, and after a few days a spray containing endrin was used. The baby and the small family dog were taken out of the house before the spraying was done about nine o'clock one morning. After the spraying the floors were washed. The baby and dog were returned to the house in midafternoon. An hour or so later the dog vomited, went into convulsions, and died. At 10 P.M. on the evening of the same day the baby also vomited, went into convulsions, and lost consciousness. After that fateful contact with endrin, this normal, healthy child became little more than a vegetable—unable to see or hear, subject to frequent muscular spasms, apparently completely cut off from contact with his surroundings. Several months of treatment in a New York hospital failed to change his condition or bring hope of change. "It is extremely doubtful," reported the attending physicians, "that any useful degree of recovery will occur." (27)

While ecological "facts" form the foundation of Carson's new vision of progress, the demonstration of these "facts" via stories lends power and urgency to her message. These stories held the attention of Carson's readers long enough for them to grasp the scientific milieu in which the stories occurred.

Science, or the search for truth, forms the background for the new terministic screen Carson was weaving. She argued that "real" science focuses on questions of truth rather than of expediency. Techniques that people construct without true understanding of natural law may produce quick, short-term results, but they have no staying power. The chemical-control advocates have replaced understanding with expediency, which guarantees disaster. Carson suggests that, while expediency may be an appropriate modifier for the goal of understanding, it must be pushed back into its supporting role.

An emphasis on biological science leads to Carson's claim that humanity is more reasonably characterized as a part of the world than as apart from the world. Carson also discusses the importance of the education function system, or the search for better solutions to problems. Because technological solutions that lack a foundation in truth are bound to fail, education is essential. Carson argues that public ignorance is no longer tolerable and urges her readers to learn for themselves, then to demand that their leaders act in society's best interest. Because nature is the best teacher, readers are validated in using their personal experiences as one basis for determining right action. Carson also urges them to question technocratic interpretations of science that rely on the slick substitution of expediency for understanding. Only by basing action on an understanding of truth (discovered by science) can humanity succeed in its quest toward a more appropriate progress.

Carson constructs a terministic screen wherein care for the environment becomes the most expedient course. When examining expediency, she incorporates society's economic, legal, and political functions. Although the economic system is not excluded from the new vision, it is integrated among other function systems rather than preeminent. A society's political function system focuses on *who* is in or out of power; its legal system focuses on ways to *use* that power to encourage valuable behaviors and to discourage those that damage the social system; and its economic system focuses on how to *pay* the costs incurred in this process (Luhmann, *Ecological Communication* 51–75, 84–93). Carson's new terministic screen relies on interaction among all three systems. She first shows that a hierarchical power structure exists and that—under the current power structure—advocates of chemical solutions to insect problems have made decisions motivated only by short-term, economic criteria. Her discussion of political power provides readers with an activist alternative to the current situation. Although readers are not personally responsible for current abuses, they are responsible for ousting those who have used power in ways that damage society. Further, they are responsible for using the legal system to discourage the indiscriminate use of pesticides. The principle economic concerns Carson displays are oriented toward procedures for achieving these goals. Economics is represented as a tool for encouraging more careful scientific study and for making the costs of legal safeguards less exacting than they otherwise might be.

The goal of progress is reconceptualized as discovery of natural truth. *Silent Spring* presses home the point that sincere truth seekers will, therefore, turn to nature in their scientific endeavors. Humans are urged to utilize all available tools to act upon what they learn from nature, the best teacher. The natural world takes on a religious transcendence as the best source of all true knowledge. Rigorously pursued, science enables people to achieve the understanding necessary to live well. Through it, humanity can learn to work with, rather than against, other life-forms.

Scientific Ecology after *Silent Spring*

One indication of whether Carson's redefinition of progress has been integrated into society is how her perspective is incorporated into university-level ecology and environmental-science textbooks. There are at least two reasons to search for Carson's influence in these texts. First, in *Silent Spring,* Carson gives a central position to science (the search for truth), emphasizing environmental sciences, such as ecology. Second, she argues that nature is the most appropriate teacher of truth. Because ecology and environmental-science textbooks are the primary vehicles for teaching university students and, indirectly, primary and secondary school students about nature, evaluation of these texts seems particularly relevant. Toward this end, we analyze five ecology and environmental-science textbooks to see whether, in the sections addressing pesticides and other environmental contaminants, the authors share Carson's terministic screen toward human-induced environmental degradation and her redefinition of progress.

Odum's *Fundamentals of Ecology*

We begin our analysis with Eugene Odum's classic textbook *Fundamentals of Ecology*. The first two editions were published prior to *Silent Spring* (1953 and 1959), the third in 1971. These books were part of the paradigm shift in ecology that led to the incorporation of a process-function perspective of ecological systems (Peterson and Peterson, "Ecology"). They presented this "new ecology" to a mass audience and, directly or indirectly, are largely responsible for the public's current understanding of ecosystem ecology.

A new terministic screen. In his first edition of *Fundamentals of Ecology,* Odum evidences an economically defined, utilitarian perspective toward the natural world, at least as it relates to indiscriminate insecticide use. He describes "injurious insects" as "man's chief competitors" (319) and, when discussing the effects of broad-scale DDT application, warns that "useful birds and fish" will be destroyed (320). He argues that although the chemical control of insects is "primarily an entomological and chemical problem," it "also enters the realm of ecology since organisms other than the intended victim may be affected" (319). Hence,

> [T]he ecologist must sometimes put a damper on the enthusiasm of the chemist and the chemical engineer who can synthesize new poisons and develop effective methods of application

faster than the total effects in nature can be determined. This is especially true when poisons are to be used in complex eco-systems such as orchards, forests, and marshes. In a number of well established cases the use of insecticides has "backfired," because of the destruction also of useful organisms such as honeybees (which pollinate many fruit trees and crops) and beneficial insect parasites. (319)

Thus, ecologists must ensure that chemical pesticides do not damage organisms that humans find useful by separating their science from pesticide technology.

In his second edition, Odum more than doubled the length of the section addressing pesticides. The new material stresses the ir-rationality of applying insecticides to complex ecosystems, such as forests, "without any knowledge of the effect the poison might have on natural control mechanisms" (425). Moreover, using chemicals to reduce the numbers of one pest often leads to increased "num-bers of another pest which is more resistant to the insecticide" (425). Similarly, "[I]n addition to the killing of beneficial organ-isms, excessive use of insecticides . . . [creates] other problems such as toxic residues in soil and human foods, off flavors in foods, and the development of resistance by pests themselves" (426). Odum argues that, because "chemical control of pests cannot be the only answer" to the pest problem (426), biological control must be given more emphasis. He then discusses techniques that subsequently were labeled "integrated pest management" as the wave of the future.

Odum's terministic screen bears only superficial resemblance to that of Carson in *Silent Spring*. For example, although he discusses-troublesome insects as the "intended victim[s]" of insecticide ap-plication (2nd ed. 425), he makes no unambiguous use of the war-fare metaphors so prominent in *Silent Spring*. Similarly, while insect pests develop resistance to insecticides, Odum does not maintain, as does Carson, that insects have an immense capacity to defeat chemical efforts to control them. He also makes no reference to public perceptions of pesticide use, and he calls upon professional ecologists, rather than the general public, to protect the environment from pesticide contamination. Further, Odum uses less-evocative language when depicting the lack of pesticide specificity. While he argues that a "damper" must be placed on chemists' "enthusiasm" (425), he does not imply, as does Carson, that those responsible for promulgating pesticide technology are consciously understating the myriad risks associated with pesticide use to ensure public ignorance of pesticide dangers.

In the third edition, Odum dramatically changes his handling of environmental contamination and pollution issues, explicitly adopting the use of metaphors so common in *Silent Spring*. This material now forms two chapters, rather than a few paragraphs, and the author retains little of the language from earlier editions. Odum now presents a quite different terministic screen to his readers. Humans are engaged in "a kind of holding action" in their "war with insects and other competitors" (446). Although "most knowledgeable scientists" warned that the "campaign to eradicate the imported fire ant" was "overkill," their advice was ignored (446). The subsequent "massive onslaught" of "saturation bombing" with insecticides resulted in "aquatic and terrestrial wildlife . . . suffer[ing] grievously" but failed to "eradicate" the pest (446). Odum writes that the "war with insects" will be lost if humankind does not begin using a "mixed bag of weapons, including old-fashioned but common-sense cultural practices, judicious use of degradable . . . chemical pesticides, and greater use . . . of nature's own control methods" (446–47). In short, humankind should use the more ecologically sound "arsenal" of "integrated [pest] control" in their battles with insects (447).

Other aspects of Odum's perspective regarding pesticides now more closely resemble those of Carson as well. For example, he tells readers that persistent, broad-spectrum insecticides, such as DDT, cannot be used without "poisoning the whole ecosystem" (455). Further,

> Pesticide pollution has been greatly aggravated by unnecessary aerial spraying of entire landscapes. Other "unforeseen" problems arise because new insecticides are tested (often very superficially) at the organism level of organization and then used at the ecosystem level without further testing. Thus, even though a chemical kills insects in cages and does not kill a laboratory rat, this does not mean that it is safe to use in nature. Again . . . trouble occurs because the agricultural and commercial specialist does not know the difference between a population and an ecosystem! (446)

Ecologists' "warnings of an entomological backlash" due to this "senseless saturation of the environment with the persistent . . . broad-spectrum poisons" went unheeded by insecticide purveyors, producing "one of the world's most serious pollution problems" (445–46). All too often, "[A] misdirected Federal government mission motivated mostly by politics . . . [is] carried out against the advice of most knowledgeable scientists" (406). "Concentration levels" of these poisons "in human tissue are now high enough" that "hormone balance" could be upset and "cancer and deleterious mutations" occur, "especially if nothing is done to control and monitor the further use of these potentially hazardous chemicals" (446). Odum then explains that this "insidious" "poisoning of entire food chains was dramatically brought to public attention in 1962 by Rachel Carson's famous book, *Silent Spring*" (445–46). Odum uses his 1971 edition of *Fundamentals of Ecology* to reiterate Carson's warning, affording it the added respectability due the premier ecology textbook.

Odum also echoes Carson's call to action. He lauds the "highly

successful control of fruit flies based on detailed scientific information and judicious use of chemicals" and warns that "the mass use
of persistent, broad-spectrum poisons . . . [must be] phased out,"
and pest management must "evolve into . . . integrated [pest] control" (446). He also echoes Carson's call for reliance on true science. In fact, "eternal vigilance, study, and trained professionals are
part of the 'disorder pumpout' in the agroecosystem. There is no
'one-shot' solution, nor will there ever be one" (447). Furthermore,
"[I]nsecticides and herbicides . . . [should] be under licensed control of trained professionals, just as are drugs used to treat the human body" (448). Awareness of the problems associated with indiscriminate chemical pesticide use is only the beginning. Both
authors argue that, to solve environmental dilemmas, awareness
must be followed by in-depth understanding of natural systems followed by sustained application of this understanding.

By 1971, Odum's perspective toward nature diverged from utili-
tarianism to one remarkably similar to Carson's. Although we do
not imply that *Silent Spring* was the sole cause of this change, it was
more than coincidental since Odum explicitly references Carson's
work. At any rate, Odum's third edition follows from, and further
legitimates, Carson's alternative vision of human progress. He now
argues, as does Carson, that humanity must systematically use the
science of ecology to understand nature, including insects. If this is
done, humans can then use the natural enemies of organisms they
consider pests in control efforts. Odum's vision is that pesticide
technology should be used only in conjunction with ecologically
sustainable, integrated pest management. In essence, humanity
must work out its position within the natural world rather than do
battle with nature.

Redefining progress. Like Carson in *Silent Spring,* Odum, in his
third edition of *Fundamentals of Ecology,* undertakes the task of
redefining human progress. The most striking change in Odum's
new treatment of environmental contaminants is his urgently, activist tone. He begins the chapter entitled "Pollution and Environ-

mental Health" by stating that "'throw-away' pollutants" and the "inevitable by-products of transportation, industry, and agriculture" are "now the most important limiting factor for man" and that pollution control must be rapidly implemented to "prevent man from completely raping the earth's resources, and thereby destroying himself" (432). He argues that recognizing pollution's costs in terms of human health "will probably do more to alert egotistical and self-centered man to the rising danger than the other kind of costs, which can be too well hidden by short-term 'cost-benefit' manipulations" (433). Furthermore,

> [A]ir pollution provides the . . . signal that may well save industrialized society from extinction because: (1) it provides a clear danger signal that man must somehow soon "power down" in the concentrated use of industrial energy, (2) everyone contributes to it . . . and suffers from it, so it cannot be blamed on a convenient villain, and (3) a solution must evolve out of holistic consideration. (445)

Odum clearly rejects the notion that environmental decisions should be based solely on the economic function system because there are "costs" that cannot be measured on a scale of dollars. Consequently, holistic evaluation of environmental problems is necessary before management decisions are made and action taken.

By including the economy as only one of many function systems, Odum elevates the importance of society's other perspectives toward nature. For example, in arguing that the "weakest link in pollution abatement strategy . . . is the inadequate legal protection of environmental quality and the consumer" (442), he centralizes the *political, religious,* and *legal* function systems in his efforts to *educate* students, for,

> [U]p to now the greatest economic rewards and the strongest legal protection have been given to those who produce, build,

pollute, and exploit nature's riches. . . . Now it is obvious that at least equal rewards and protection must be given to those people, professions, and industries that maintain the quality of human existence; survival in the future depends on finding a balance between man and nature in a world of limited resources. This does not mean that man must revert to nature, but it does mean that he will need to go back to some of the good, common-sense, old-fashioned things such as returnable bottles, walking, and human concern for one's neighbors. Some things, one-way bottles for example, that we once thought represented "progress" turned out instead to be insults to both man and nature. (442–44)

This is a cogent moral and pragmatic argument for politicians to pass laws that make economic incentives (tools) available to those who maintain or improve environmental quality. It is a clarion call for humans to use ecological wisdom to save themselves and the earth around them.

Not surprisingly, Odum's new vision of progress, like Carson's, affords a central position to the science of ecology. His insistence that humans must make substantially greater use of biological control of insects they consider pests (447) is just one example of his contention that humankind must develop an ecological perception of the earth. He argues that only a reliance on sound ecological principles can give humanity a workable alternative to the continued "rape" of the earth, the eventual destruction of modern human society, and possibly the extinction of *Homo sapiens*. Odum does not, however, imply that science is the only function system needed to address environmental contamination. He makes a cogent argument, in fact, for integrating multiple systems. Like Carson, Odum maintains that a public better educated in ecology would insist that politicians enact statutes needed to protect nature (including humans) from the unscrupulous, use the courts to ensure that these laws are enforced, and utilize the economy to pay for required social changes.

Current Ecology and Environmental-Science Textbooks

We now evaluate two current ecology/environmental-science textbooks to determine whether their authors share Odum's (3rd ed.) and Carson's terministic screen and redefinition of progress. We chose G. Tyler Miller's *Living in the Environment* as an example of textbooks designed for entry-level environmental-science courses—where human perceptions of the environment, various social issues, and "calls to action" are integral parts of the curriculum. We also examined *Ecology: Individuals, Populations and Communities*, by Michael Begon, John L. Harper, and Colin R. Townsend, as an example of textbooks designed for upper-division or graduate courses in ecology—where calls to action are typically more restrained, and the primary emphasis is on presenting the science of ecology, in all its detail, to students. Although there are numerous other textbooks we could have chosen, these two are well regarded by the scientific community, commonly used, and representative of their respective genre.

A new terministic screen. These contemporary textbooks still view human interactions with insect pests in martial terms. Begon et al. explain that inorganic and botanical insecticides "were the chemical weapons of the expanding army of insect pest managers of the nineteenth and early twentieth century" and that chemical pesticides "are now used as an integral part of a more varied armoury wielded against the pest" (558). However, this integrated pest management requires "training the necessary army of advisors" (580). Similarly,

> [T]he longest war in human history is our war against insects.
> This war was declared about 10,000 years ago, when we first

got serious about agriculture, and we are no closer to winning today than we were then. In fact, the multitudes of insects that share our fondness for rice and corn, cotton and wheat, and beans and apples seem to be gaining on us. (Miller 618)

Although we may be losing this war, humans are not the only species battling pests. "Monarch butterflies, golden toads, and bombardier beetles also wage chemical warfare against their predators" (620), and plants have "been at war with insects for eons" (621). Apparently, warfare with insects is acceptable under certain conditions, but the weapons used may be problematic.

This war pits humans against a formidable enemy. After all, "'[B]ugs are not going to inherit the earth. They own it now. So we might as well make peace with the landlord.' As we seek new ways to coexist with the real rulers of the planet, we would do well to be sure spiders are in our corner" (Miller 619). In the 1950s, the public was led to believe that chemical insecticides were the weapons that would finally enable humankind to win this war. However, "[T]he publication of Rachel Carson's *Silent Spring* in 1962 . . . heralded a public challenge to the pervasive notion that the benefit/risk equation for the use of these chemicals was tilted overwhelmingly in favour of benefits" (Begon et al. 557). Finally, more than "three decades after Rachel Carson's warning, we are beginning to get the message that trying to eradicate insects and other pests with massive chemical warfare won't work because insects outnumber us, outbreed us, and rapidly develop . . . [resistance] to the chemicals we throw at them" (Miller 624).

Tragically, our officers in the war against pests betrayed our trust. Pesticide technologists made "unthinking use of chemical pesticides in the 1940s and 1950s" (Begon et al. 579), and insecticides were often "applied carelessly." "Devastation" of our allies—untargeted insects, livestock, cats, dogs, and wildlife—occurred due to this friendly fire (562). To make matters worse, insecticides often kill the "natural enemies" of the insect pest, which "may not seem too serious apart from the regrettable loss in natural diversity of harmless species" (564):

> However, this can—and often has had—two extremely serious consequences. The first, target pest resurgence, . . . occurs when treatment kills not only large numbers of the pest, but large numbers of their natural enemies too (with survivors likely to starve to death because there are insufficient pests on which to feed). Then, any pest individuals that survive . . . find themselves with a plentiful food resource but few if any natural enemies. A population explosion is the likely outcome. . . . [because] the enemies need the pest to support population growth, but the pest certainly does not need the enemies. (564)

The second serious consequence is that many "non-pests become pests when their enemies and competitors are killed" (565). Although humankind may need to battle pests at times, misguided barrages of chemical weapons only abet the enemy.

The symbolic reality humanity constructs regarding pests is a large part of the pest problem. After all, because too many people think that "the only good bug is a dead bug" (Miller 633), "the average U.S. homeowner applies three to six times more pesticide per hectare than do farmers" (622). Furthermore, because "humans are a demanding lot" (Begon et al. 551), our "insistence on perfect, unblemished fruits and vegetables, even though a few holes or frayed leaves do not significantly affect the taste, nutrition, or shelf life of such produce," and our "obsession with the perfect lawn" force us onto "the pesticide treadmill" (Miller 633). To get off this treadmill, humans must redefine progress such that everyone "recognize[s] that pest control is basically an ecological, not a chemical, problem" (634). In sum, both texts argue that society must rely on ecologically rather than chemically based integrated pest management because this is the only sustainable solution to our pest problems. This description of progressive society could as easily have been written by Carson.

Redefining progress. Miller and Begon et al. echo Carson's and Odum's (3rd ed.) calls for a revised definition of progress that centralizes science (ecology). Economics, rather than being the central figure, recedes as one among many societal functions. After all, this is the only way humans can maintain peace with species we consider pests and become one with nature rather than fight against it. These authors argue that humanity's political, legal, educational, and economic perceptions of nature can be productively integrated to achieve this morally transcendent, yet pragmatic, goal.

Begon et al. offer examples of politically and legally based efforts toward this new progressivism. Politicians in the United States passed laws that outlawed the "notorious" DDT "from all but emergency uses" (559), and lawsuits led the manufactures of Agent Orange to compensate U.S. Vietnam War veterans in an out-of-court settlement, for various maladies, including birth defects and certain rare cancers apparently caused by their product (564). Politicians also formed the Environmental Protection Agency (EPA) and banned several hazardous pesticides, including most chlorinated hydrocarbon insecticides, several carbamates and organophosphates, and the herbicides 2,4,5-T and Silvex (Miller 627). Unfortunately, the EPA and the Food and Drug Administration still "inadequately and poorly" enforce federal laws regulating the use of pesticides in the United States (627). For this reason, environmentalists use a 1988 statutory provision to bring "citizen suits against the EPA for not enforcing the law, an essential tool to ensure governmental compliance" with statutes (628). Worse yet, politicians left "loopholes" in the law, allowing "the EPA to leave inadequately tested pesticides on the market and to license new chemicals without full health and safety data" (627–28). Similarly, even though the U.S. Congress banned the use of certain dangerous pesticides in the United States, they refuse to break the "circle of poison" by banning the export of these same chemicals and still "allow food with residues of those pesticides to be imported for U.S. consumers" (628). Although the political and legal function systems have successfully been used to promote ecologically sound interactions with the environment, much political and legal work remains.

Unlike Odum (3rd ed.), Miller and Begon et al. do not claim that economic arguments are inadequate, and sometimes inappropriate, as justifications for using chemical insecticides. Rather, they argue that more ecologically sustainable alternatives, such as integrated pest management, also are more economically viable. After all, "biological control is cheap" (Begon et al. 567) and saves "farmers an average of $25 for every $1 invested" (Miller 629). Unfortunately, integrated pest management "is hindered by government subsidies of conventional chemical pesticides and by [the] opposition of agricultural chemical companies" (632). Hence environmentalists urge politicians to pass laws

> adding a 2% sales tax on pesticides to fund IPM [integrated pest management] research and education; setting up a federally supported demonstration IPM project on at least one farm in every county; training USDA field personnel and county farm agents in IPM so they can help farmers use this alternative; providing federal and state subsidies and perhaps government-backed crop-loss insurance to farmers who use IPM or other approved alternatives to pesticides; [and] gradually phasing out subsidies to farmers depending almost entirely on pesticides once effective IPM methods have been developed for major pest species. (633)

This reinterpretation of progress relies upon *politicians* passing *laws* making *educational* and *economic* incentives available that encourage farmers to use ecologically (*scientifically*) sustainable alternatives to pesticides. Thus, these authors echo Carson's and Odum's (3rd ed.) calls not only for integrated pest management but also for a reintegration of society's various function systems so that the science of ecology is afforded a more central position.

Ecological Progress

Silent Spring offers a new progressive paradigm that fosters care for the environment, awareness that nature provides the foundation and, therefore, the appropriate boundaries for all technological innovation, and an assumption that the quality of human life depends on sustaining that foundation. Carson argues that chemical warfare is not an effective means for controlling insect pests. More important, she claims that the decision to wage war against the rest of nature is unhealthy for humanity. She advocates a radical reordering of priorities that encourages humans to learn how to live with the rest of nature rather than how to fight against it.

Burke claims that hierarchy is intrinsic to social systems, and Carson does not escape the hierarchical principle. But the apex of her hierarchy is science rather than economics. Her radical reordering of priorities does not leave the reader without alternatives, however. *Silent Spring* offers a terministic screen that promotes awareness of ambiguity and irony. Carson uses new rhetorical combinations to transcend the categorical and logical barriers constructed by society's traditional terministic screen and rigidified by its technological dependence. She promotes integrative, social knowledge wherein people are corrected rather than punished for past mistakes. She contextualizes all this in an approach that encourages the reader to think of society as multifaceted. Short-term economic benefits are important but are not the sole criteria for decision making. The ability to pay for things should not mandate all decisions. The incongruous perspective she offers is the claim that human society will fare better if we understand ourselves as belonging within nature rather than positioned adversarially toward other life.

Silent Spring encourages readers to reinterpret their relations with the nonhuman world by wrenching technology from its privileged position at the right hand of science. Carson highlights at least two fundamental errors made by technocrats. First, they have made enemies of insects that might have been their allies, and certainly need not be enemies. Second, they have failed to take account of the natural abilities possessed by those enemies. They have structured a war in which insects have no choice but to destroy or be destroyed. Carson recasts those technocrats who have played the role models for rational progress as "primitive . . . cave men" (297). She redefines them as "fanatics" (12) who deal in "myth" (9) and "legend" (34) in page after page of anecdotes, scientific reports, and cogent arguments. She strips from them the mantle of science, describing them as mentally "narrow" (63), "authoritarian" (127) purveyors of "illusion" (114). She writes that the unnecessary suffering they visit on all life "in the name of progress" (66) is the "height of absurdity" (158). By removing technology from its leadership role, Carson leaves room for a new vision of progress.

Silent Spring offers an analytical frame for discovering aspects of our social constructs that needed revising, and it invites its readers to begin that project. The authors of the ecology and environmental-science textbooks we have examined have accepted this invitation. Odum's third edition of *Fundamentals of Ecology,* for example, diverges from the utilitarian perspective toward nature that characterized his earlier editions to one where ecologically sound methods of pest management are not only pragmatic but also morally imperative. Similarly, both Miller and Begon et al. expand upon Carson's call to implement biological and cultural control of pests by demonstrating the necessity for ecologically, rather than chemically, based pest management. Like Carson, they maintain that this is the only way to coexist with the rest of the natural world and be good citizens of Earth. The authors share Carson's terministic screen, where progress is reinterpreted so that ecological science takes a more central role in society.

Further, Odum (3rd ed.), Miller, and Begon et al. understand society as a multifunctional system where communication across function systems is the only means whereby humanity can respond appropriately to environmental complexity. Although all three texts reinterpret progress so that society's function systems are reintegrated, with the science of ecology centrally positioned, this does not imply that other systems are unimportant. Rather, the political, legal, educational, and economic systems form integral and necessary aspects of the new vision of progress. Humanity should, they argue, systematically use ecological science to understand nature and only use pest-management approaches that are ecologically sustainable. There are, however, many economic and noneconomic "costs" of pollution to humanity's and other organisms' health and well-being. While the authors decentralize the importance of the economic function system, they do not imply it is insignificant. In fact, they all state that responsible pest management can be less economically expensive in the long term than remaining on the pesticide "treadmill." They envision human society as a subsystem within the larger biosphere that can increase society's ability to respond to environmental perturbations by increasing the resonance between function systems. No single function system possesses a sufficiently complex terministic screen to account for all reality. By facilitating communication across functions, a society can exponentially increase the types of reality that can become information, thus diversifying options for interacting with nonhuman life. These authors' ideal society places science at the crossroads, guiding communication among all other function systems. Although they lament that so little progress has been made toward implementing this vision for the future, they clearly share the essence of *Silent Spring*'s message.

Opie and Elliot write that Carson's "errand was to stop the scientific management of nature" (34). Although this may be true of "A Fable for Tomorrow," taken as a whole, *Silent Spring* is an attempt to revitalize a refined scientific understanding of nature, thus seeking to stop the technological domination of science. In repositioning humanity as part of, rather than apart from, nature, Carson offers a new vision of science that emphasizes human affinity with other living creatures rather than with technologically constructed objects. Within such a universe, science focuses on understanding rather than control. She suggests that a more holistic understanding of nature's complexity will enable us to decide intelligently which aspects of nature we should attempt to manage and which are better left alone. Further, holistic understanding provides the best foundation for deciding how to control those aspects of nature we must manage. When Carson writes that "life is a miracle beyond our comprehension, and we should reverence it," she adds, "even where we have to struggle against it" (275). This is not the writing of a naive nature child. The book's entire concluding chapter discusses the "extraordinary variety of alternatives to the chemical control of insects" (278) that demonstrate "the worth of scientific creativity, aided by thorough basic research, persistence, and determination" (281). In all of these cases, however, "[N]ature has pointed the way" (285). *Silent Spring*'s new vision of progress pictures a scientifically enlightened humanity, uncoupled from its technological handicap, "cautiously seeking to guide [living creatures] into channels favorable to ourselves" (296).

Works Cited

Begon, Michael, John L. Harper, and Colin R. Townsend. *Ecology: Individuals, Populations and Communities.* 2nd ed. Boston: Blackwell, 1990.

Burke, Kenneth. *Attitudes Toward History.* 1937. 3rd ed. Berkeley: U of California P, 1984.

———. *Language as Symbolic Action: Essays on Life, Literature, and Method.* Berkeley: U of California P, 1966.

———. *Permanence and Change: An Anatomy of Purpose.* 1935. 3rd ed. Berkeley: U of California P, 1984.

————. *A Rhetoric of Motives.* 1950. Berkeley: U of California P, 1969.

Cantrill, James G., and Christine L. Oravec, eds. *The Symbolic Earth: Discourse and Our Creation of the Environment.* Lexington: UP of Kentucky, 1996.

Carson, Rachel. *Silent Spring.* 25th anniv. ed. Foreword by Paul Brooks. Boston: Houghton, 1987.

Killingsworth, M. Jimmie, and Jacqueline S. Palmer. "Liberal and Pragmatic Trends in the Discourse of Green Consumerism." Cantrill and Oravec 219–40.

————. "Millennial Ecology: The Apocalyptic Narrative from *Silent Spring* to *Global Warming.*" *Green Culture: Environmental Rhetoric in Contemporary America.* Ed. Carl G. Herndl and Stuart C. Brown. Madison: U of Wisconsin P, 1996. 21–45.

Luhmann, Niklas. *Ecological Communication.* Trans. John Bednarz, Jr. Chicago: U of Chicago P, 1989.

————. "What Is Communication?" Trans. John Bednarz, Jr. *Communication Theory* 2 (1992): 251–59.

Miller, G. Tyler, Jr. *Living in the Environment: Principles, Connections, and Solutions.* Belmont, CA: Wadsworth, 1994.

Odum, Eugene P. *Fundamentals of Ecology.* Philadelphia: Saunders, 1953.

————. *Fundamentals of Ecology.* 2nd ed. Philadelphia: Saunders, 1959.

————. *Fundamentals of Ecology.* 3rd ed. Philadelphia: Saunders, 1971.

Opie, John, and Norbert Elliot. "Tracking the Elusive Jeremiad: The Rhetorical Character of American Environmental Discourse." Cantrill and Oravec 9–37.

Oravec, Christine L. "To Stand Outside Oneself: The Sublime in the Discourse of Natural Scenery." Cantrill and Oravec 58–75.

Peterson, Markus J., and Tarla Rai Peterson. "Ecology: Scientific, Deep and Feminist." *Environmental Values* 5 (1996): 123–46.

Peterson, Tarla Rai. *Sharing the Earth: The Rhetoric of Sustainable Development.* Columbia: U of South Carolina P, 1997.

Peterson, Tarla Rai, and Markus J. Peterson. "Valuation Analysis in Environmental Policy Making: How Economic Models Limit Possibilities for Environmental Advocacy." Cantrill and Oravec 198–218.

6

CAROL B. GARTNER

When Science Writing Becomes Literary Art: The Success of *Silent Spring*

Rachel Carson's major achievement in *Silent Spring* is a fusion of science and literary art so seamless that the effect is seductive. To her contemporary critics in industry and science, it was so effective it was insidious. Her basic strategy was pragmatic, based upon a meliorist philosophy: give people knowledge in a form they can understand, and they will act on it; show people how we are destroying our earth, and they will move to curb the destruction. Her goal was to initiate change. Never did she use "art for art's sake" or seek merely to dazzle or delight. She followed the classical approach to rhetoric: to please *and* to teach. It was because of her books' literary qualities that vast numbers of people eagerly read

This essay is a revised and expanded version of portions of "A Book for Our Time: *Silent Spring*," chap. 6 of Carol B. Gartner's book *Rachel Carson*, New York: Frederick Ungar Publishing Co., 1983, 86–109. Copyright © 1983 Frederick Ungar Publishing Co., reverted 1998 to Carol B. Gartner. Used by permission of the author.

them. Science is the content, but art enhances its communication and multiplies the persuasiveness of both the scientific argument and the ecological philosophy that underlies it. When Carson received a National Book Award for *The Sea Around Us*, she told her audience that there is "no separate literature of science." "The aim of science," she said, "is to discover and illuminate truth. And that, I take it, is the aim of literature" (qtd. in Brooks 128).

Carson believed that content must determine structure, the concept known in poetry as *organic form*. "The subject takes command," she wrote, "and the true act of creation begins" (qtd. in Brooks 2).[1] She structured *Silent Spring* as a legal argument, as logical, pointed, and convincing as a good law case, with scientific research citations for every point. In each section, the "case" is closely argued with details and proofs, including an array of corroborating experts. Carson proposes a specific remedy for every area of damage she describes. The book as a whole is tightly organized through the use of recurrent themes, motifs, and images to provide structural interrelationships and overall unity.

Because of the meticulous accuracy and rigor with which Carson supported her claims, much of the science she presented is still sound. As in *The Sea Around Us*, the general framework and propositions remain valid, even where later researchers have modified the specifics or reinterpreted the data. Knowledge of the factors involved in the development of human cancers has increased dramatically since the publication of *Silent Spring*, but Carson's basic presentation of theories of the origin of cancer cells remains remarkably accurate. Critics greeted the book with many accusations of overstatement and misinterpretation of facts. In all the many and varied scientific discussions that fill *Silent Spring*, however, the discussion of cancer statistics is the only place where she may have drawn implications not completely supported by the facts she presents.

Carson uses a lawyer's arsenal of classic rhetorical argumentation to make her case but augments it with a writer's mastery of poetic technique. The beauty of her writing beguiles the reader into reading and assimilating material that is both intellectually difficult to understand and emotionally difficult to accept. Her literary skill is most evident in her ability to present complex scientific information with both clarity and grace. A graphic simile, for example, makes it possible for the reader to picture an intracellular process in the guise of a commonly known machine.

Finally, Carson demonstrates her mastery of the rhetorical principle of audience analysis—a skill crucial to any literary artist—through pragmatic appeals to human self-interest interwoven with her calls for a more altruistic concern for preservation of the natural world. Only gradually does she push her readers toward the more difficult concept that preserving the greater environment is implicit in preserving ourselves. Unlike some more extreme later ecologists, who found in her work part of the inspiration for their movement, Carson believed a knowledgeable and aroused public could bring about immediate changes in government policy and industry practice, beginning what she realized would be a long, incremental process. She shaped her strategies in *Silent Spring* to reach the general public and mobilize what she knew could be formidable forces to counter industry and to insist upon change.

The Case for the Prosecution

Carson went to the court of public opinion to accuse the chemical and agricultural interests, and the government officials who abetted them, of committing a series of crimes. She wanted not to punish them but to get them to mend their ways. For example, she wanted them to stop pursuing short-term gains, such as increasing crop yields through the use of pesticides and other chemicals, that would have disastrous long-term consequences. She presents her exposé of thoughtless practices by both industry and government as a powerful "case," making her intentions clear from the beginning. Chapter 2, "The Obligation to Endure," is the opening argument for the prosecution, with its use of "I do contend" and, again, "I contend" (12–13). She continues with a series of chapters structured to include propositions, charges, details, and proofs. In chapter 7, "Needless Havoc," she discusses attempts to eradicate the Japanese beetle in Illinois, maintaining that authorities (1) do not look at history, citing successful use of natural controls in the East as proof; (2) are shortsighted and misrepresent facts, citing specifics about the cost of using milky spore disease as a natural control and arguing that it has to be done only once, whereas spraying must be repeated; (3) do not weigh potential disadvantages, such as extraneous destruction and possible harm to humans; and (4) look for "immediate results," even if they do not last (91–99). A syndicated review of the book declared that "she presents her evidence like a public prosecutor, with a relentless battery of testimony" (Smith).

Except for the opening "Fable," *Silent Spring* is densely packed argument, moving from the exhortation to endure through chapter after chapter of problems and cogent suggested solutions. The many interrelated ideas—central to one topic but peripheral to others—and the repeated examples, such as the fire ant and Japanese beetle programs, brought formidable organizational challenges. Carson imposed strict logical development and used recurrent themes, motifs, and images to achieve structural interrelationships and overall unity. In the "Fable," for example, she introduces the killing of birds, wildflowers, and fish. These deaths become symptoms and ultimately symbols of the inadvertent side effects of pesticide use.

Chapter titles are echoed in the text. Chapter 2, "The Obligation to Endure," ends, "In the words of Jean Rostand, 'The obligation to endure gives us the right to know'" (13). The phrase "present road" (13), used in the second chapter to describe our current actions, foreshadows the final chapter, "The Other Road," which presents our other options. Neighboring chapters are similarly connected. At the beginning of her chapter on the soil, Carson uses the phrase "the earth's green mantle" (53), which is the title of the following chapter. That chapter opens thus: "Water, soil, and the earth's green mantle of plants make up the world that supports the animal life of the earth" (63). The sly last sentence in chapter 11, "As matters stand now, we are in little better position than the guests of the Borgias" (184), refers not only to the title, "Beyond the Dreams of the Borgias," but also to the thematic use of the Borgias, benevolent on the surface, as prototypical poisoners. Earlier in the book, she describes arsenic as an agent of homicide "from long before the time of the Borgias" (17).

Not only must a good case be effectively organized, but it must convincingly establish the validity of its facts and its interpretations. Carson did this through meticulous documentation and corroboration by experts in every field. Her collected papers demonstrate her exhaustive preparation for writing *Silent Spring*. There are collections of scientific reprints, conference reports, congressional testimony, newspaper articles, and letters ranging from expert testimony in answer to her technical questions to personal accounts of illness or observations (Rachel Carson Papers). Anticipating controversy, Carson included fifty-five pages of notes documenting her sources in *Silent Spring* (301–55). She calls them to the reader's attention with a disarmingly gentle author's note just before the acknowledgments. References throughout the text cite specific authorities, and the acknowledgments include sixteen experts who had read and commented on portions of the manuscript.

Science That Endures

Thanks to Carson's meticulous research and her talent for keen scientific insight and synthesis, *Silent Spring* retains remarkable scientific reliability. The basic ideas and scientific principles are still valid, even in fields like cancer research where the expansion of knowledge in the years since 1962 has been vast and the understanding of the mechanisms of cellular change, based on genetic research, has increased greatly. In chapter 14, "One in Every Four," Carson discusses two theories of the origin of cancer cells: the genetic theory that cancer results from chromosome damage; and German scientist Otto Warburg's theory that the respiration of normal cells is destroyed, throwing them back on a less-efficient, primitive method of energy production, forcing repeated cell division to compensate.

As to the second theory, Drs. Elizabeth and James Miller noted in 1981 that newly developed sophisticated techniques to measure energy within the cell made it possible to investigate the validity of Warburg's theory (1055). In a 1993 article, we find a historical review of the dominant theories of cancer production, beginning in 1926 with Warburg. Discussing all the different theories, the authors

> point out that these approaches are by no means mutually exclusive and, even if they are not generalizable, they may still be valid for specific types of cancer. Nor can it be asserted that current knowledge about the genetics of cancer has rendered these concepts obsolete. . . . Although cellular anoxia [Warburg's theory] is not a cause of cancer, it is a very common consequence of neoplastic change. Chromosomal changes and point mutations [the genetic theory] are crucial events in carcinogenesis and lie at the heart of all current genomic models. (Goldberg and Diamandis 2366)

The authors stress, however, that the "genotype of cancer is complex and confusing" and that "the phenotypic behavior and properties of cancer cells are much more so" (2360).

Scientists now focus on the genetic theory, expressed in terms of changes in the genome. The Vogelstein–Lane model of a multistep process is most current (Goldberg and Diamandis 2367; Rowley 1996). Despite significant gains in understanding the genetic model, Carson's clear description of the theory is generally valid:

> According to the mutation theory of the origin of cancer, a cell, perhaps under the influence of radiation or of a chemical, develops a mutation that allows it to escape the controls the body normally asserts over cell division. It is therefore able to multiply in a wild and unregulated manner. The new cells resulting from these divisions have the same ability to escape control. (233)

The fundamental recent knowledge of chemical carcinogenesis comes from the Drs. Miller (Goldberg and Diamandis 2366; Rowley 1981). First described as having two stages, the process is now seen as having three distinct stages. The first is initiation—"one or more mutations of cellular DNA" (Miller and Miller 1055)—which is irreversible. "The resulting cell has the potential to develop into a clone of neoplastic cells but will not necessarily do so." The second stage is promotion, which is reversible. The dose of the initiating agent is critical. The final stage is progression, which is irreversible and "accompanied by clearly recognizable genomic alterations" and evidence of discrete tumors. "Complete carcinogens" can bring about all three stages, while others may be only initiating agents (Goldberg and Diamandis 2366–67). Despite the current acceptance of the genetic theory, the role of replication, or cell division, prominent in Warburg's theory, appears in recent discussions of the use of data from animal studies in predicting risks to humans from environmental chemicals that do not themselves cause DNA mutation (Croy; Health Council).

Carson was aware of the concepts of initiation and promotion and of the interaction of chemicals to compound the damage (potentiation). Summarizing a discussion of herbicides and skin tumors, she writes: "Cancer may sometimes require the complementary action of two chemicals, one of which sensitizes the cell or tissue so that it may later, under the action of another or promoting agent, develop true malignancy" (238). She repeatedly stresses our need for further knowledge of the effects of chemicals. Her discussion of the effects of DDT on the liver concludes: "No one yet knows what the ultimate consequences may be" (23). In his authoritative discussion of the effects of environmental pollution on the liver, Dr. Hyman Zimmerman commented in 1978 that "the approach to evaluation of the character, extent, and biologic impact of chemical contamination of the environment has been rather random, and the available data are grossly insufficient to assess the effects of pollution on the liver" (334–35). Zimmerman urged that there be further investigation. He credited Carson with opening "the era of concerned focus on the problem" (333).

Carson strongly believed that DDT would be proven harmful to human beings, but this has not yet been borne out by research. Zimmerman cites proven liver damage in animals; but as for humans, he states, "Despite the billions of pounds of DDT that have been manufactured and used, no instance of hepatic injury acquired as the result of occupational or environmental exposure has yet been attributed to it." He concludes, however:

> Reassuring as is the lack of overt hepatic injury from DDT on ordinary exposure, . . . the resolution of the long-term threat of this agent to the liver is more complicated. . . . Evidence of accumulation of DDT in humans is impressive. . . . Evidence that accumulation . . . has an adverse effect on health, however, is less compelling. . . . The essential epidemiologic studies . . . remain to be performed. (336–37)

He substantiates Carson's indictments of other organic pesticides, such as aldrin and dieldrin, as proven dangerous and suggests that the most important potentially adverse effect of all contaminating chemicals may be their alteration of the function of enzyme systems, changing responses to other agents, and perhaps strengthening the potential of other chemicals to damage the liver or cause cancer (337–41).

Although scientists today generally believe that DDT does not have adverse effects on human beings, current investigations of the complex multistep process of chemical carcinogenesis suggest varied possibilities for the interaction of causative or promoting factors, as Zimmerman believed would be the case. All carcinogens may not be mutagens. In a discussion of "genotoxic and non-genotoxic chemically induced carcinogenesis," Ashby noted in 1995 that the basic question is "whether the process of carcinogenesis was initiated by the test chemical damaging DNA" (209–10). That is, was there a prior initiating chemical, allowing the test chemical to bring about the next stage of the process? In a 1993 publication, a table entitled "Chemicals and Cancer in Humans: Carcinogenicity First Observed in Experimental Animals and Subsequently by Epidemiologic Evidence" includes "DDT and related compounds" among "chemicals causally or probably associated with cancer in humans" (Huff 20).

When *Silent Spring* was published, many critics attacked both Carson's science and the implications and conclusions she drew from her facts. After commenting that she was "a biologist highly respected for an earlier book 'The Sea Around Us'," the reviewer for the London *Economist* accused her of making "propaganda play" with cancer statistics (248). Unlike their other charges, such as "unscientific use of italicized innuendo" and "disregard for the studies of the problem by her fellow-scientists in industry, the universities and government service" (251), this one has some validity, but the discussion of cancer statistics is the only such example of "playing" with facts in the book. Carson implies that chemicals in the environment have caused an increase in malignant disease but offers proof only for the increase, not for the causal connection, and even this proof is based on oversimplification of the facts:

> No longer are exposures to dangerous chemicals occupational alone; they have entered the environment of everyone—even of children as yet unborn. It is hardly surprising, therefore, that we are now aware of an alarming increase in malignant disease.
>
> The increase itself is no mere matter of subjective impressions. The monthly report of the Office of Vital Statistics for July 1959 states that malignant growths, including those of the lymphatic and blood-forming tissues, accounted for 15 per cent of the deaths in 1958 compared with only 4 per cent in 1900. (221)

"Alarming increase" was an overstatement. Even now scientists are not certain whether there has been an increase in the incidence of cancer or in deaths from cancer, even though the percentage of deaths in the United States due to cancer has continued to rise. The situation is different for different parts of the body, as well as for different portions of the population. Lung cancer is increasing, most recently in women, but its close association with smoking has been proven. Non-Hodgkin's lymphoma, testicular cancer, and leukemia are increasing, but stomach cancer is decreasing. In addition, epidemiologists have changed their statistical base to better take into account underrepresented groups in the population, such as African Americans. All combine to make comparisons difficult (information from Rowley; Schottenfeld; Coleman et al.; and Holleb et al.).

Other factors confused the statistical picture in Carson's time—and still do. The incidence of cancer rises significantly after age sixty-five, and there has been a considerable increase in the number of older people in the population. In addition, diagnostic techniques and reporting methods have steadily improved. These elements contribute to higher figures for both the incidence of cancer and cancer deaths. As for Carson's implication that environmental pollutants cause cancer, however, the evidence is clear. Many are proven carcinogens.

Science as Art

Carson's ability to present complex scientific material with beauty as well as clarity greatly contributes to the effectiveness of *Silent Spring*. Some passages may remind us of Carson's earlier books—her superb description of the formation of soil and the interaction of creatures within it, for example, recalls descriptions of the minute denizens of the sea—but the expository demands in *Silent Spring* are far greater. Carson must explain complex material to nonscientists. A prime example is the chemical structure of chlorinated hydrocarbons and organic phosphorus compounds, the two large groups of modern insecticides. All of these compounds are built on the basis of carbon atoms, the building blocks of the living world, which is why they can interfere with the processes of life. Carson must establish how much important information is as yet unknown—such as the long-range biological effects of insecticides. She must carefully delineate the practical and financial advantages of nonchemical alternatives. She must introduce broad ecological concepts, such as "the conservation of variety," which prevents the easy spread of disease that occurs in large areas planted with a single species (117).

Carson's educational method gracefully and seamlessly integrates technical explanations, broad discussions, narrative examples, and personal accounts from her own experience and those of others. Frequent changes of pace and levels of difficulty help to keep the reader's attention. Carson even slips in practical advice:

insecticide residues on food "are little affected by washing—the only remedy is to remove and discard all outside leaves of such vegetables as lettuce or cabbage, to peel fruit and to use no skins or outer covering whatever. Cooking does not destroy residues" (179). Summaries enforce our learning process and add the poetic lilt of a refrain: "The robins, then, are only one part of the chain of devastation linked to the spraying of the elms, even as the elm program is only one of the multitudinous spray programs that cover our land with poisons" (109); or, health problems—"born of the never-ending stream of chemicals of which pesticides are a part, chemicals now pervading the world in which we live, acting upon us directly and indirectly, separately and collectively" (188).

The contrast between Rachel Carson's style and that used in most scientific writing, even when intended for the general public, helps explain her distinctive power. She begins each chapter with a strong thesis, which she gradually elaborates, moving with tight organization from general statements to supporting statements to specific details. "Elixirs of Death" begins, "For the first time in the history of the world, every human being is now subjected to contact with dangerous chemicals, from the moment of conception until death" (15). This opening sentence tells us she will cover the historical development of chemicals, their nature and dangerous effects, and their ubiquitous presence. A general statement about "simpler inorganic insecticides of prewar days" is followed by general examples, such as "compounds of arsenic, copper," and then particular examples, like "pyrethrum from the dried flowers of chrysanthemums." Like all poets of the people, Carson draws on common experience for her examples, suggesting that "the common salad bowl may easily present a combination of organic phosphate insecticides" (32), and forcing readers to consider whether the threat of genetic damage is "not too high a price to pay for a sproutless potato or a mosquitoless patio" (216).

She uses the connotations of words, as well as their definitions, to build tone and ultimate meaning. Even "man's inventive mind," representing the creativity of the scientist, becomes sinister when followed by the word *brewed,* suggesting witchcraft and mad scientists:

> The chemicals to which life is asked to make its adjustment are no longer merely the calcium and silica and copper and all the rest of the minerals washed out of the rocks and carried in rivers to the sea; they are the synthetic creations of man's inventive mind, brewed in his laboratories, and having no counterparts in nature. (7)

Such verbal associations help support Carson's contention that "progress" in insect control has backfired, that to battle against nature is unnatural. In the final chapter of the book, she repeats the phrase "man's inventive mind" without negative connotations, as she describes the beginning of "The Other Road"—the use of natural or "biotic control of insects" (288)—to leave us with a positive alternative.

Carson establishes tone not only through word choice but also by using leading rhetorical questions, often tinged with irony or even sarcasm:

> The bitter upland plains, the purple wastes of sage, the wild, swift antelope, and the grouse are then a natural system in perfect balance. Are? The verb must be changed—at least in those already vast and growing areas where man is attempting to improve on nature's way. . . . to satisfy the insatiable demands of the cattlemen. (66)

The bitter tone and sometimes extravagant word choice—*insatiable,* for example—led to charges of emotionalism and sensationalism. That Carson presented her material with scientific accuracy has been repeatedly validated, but she does use her own emotions—anger, bitterness—and occasional sensationalism to develop tone and thus appeal to the reader's emotions as well as intellect. She creates an ironic and angry oxymoron, "Elixirs of Death," for the title of her third chapter, paradoxically linking *elixir,* meaning "life-giving potion" or "cure-all," with *death.*

Although she is not often humorous, she sometimes moves from irony into sarcasm: in England, "[T]he Ministry of Agriculture considered it necessary to give warning of the hazard of going into the arsenic-sprayed fields, but the warning was not understood by the cattle (nor, we must assume, by the wild animals and birds)" (35). To the author of a paper on chemical destruction of wildflowers along roadsides, she tells us, "[M]any of us would unquestionably be suspect, convicted of some deep perversion of character because we prefer the sight of the vetch and the clover and the wood lily . . . to that of roadsides scorched as by fire" (72).

When Carson writes of "a world that is urged to beat its plowshares into spray guns" (69), she introduces a sardonic play on words reversing the biblical injunction to beat swords into plowshares. The association of pesticides with warfare becomes a thematic extended metaphor. When it becomes an article of faith, no longer subject to logical argument, it is a "crusade." Thoughtless use is primitive warfare. In order of appearance, Carson characterizes this war as follows:

1. Thus the chemical war is never won, and all life is caught in its violent crossfire. (8)
2. The crusade to create a chemically sterile, insect-free world seems to have engendered a fanatic zeal on the part of many specialists and most of the so-called control agencies. (12)
3. All these [insects] have been our allies in keeping the balance of nature tilted in our favor. Yet we have turned our artillery against our friends. (251)

4. As crude a weapon as the cave man's club, the chemical barrage has been hurled against the fabric of life. (297)

Throughout *Silent Spring*, Carson uses the war motif with negative
associations; but when, at the end, she wants to present biological
controls as a viable alternative, she draws positively on the metaphor of war: "Some of the most interesting of the recent work is
concerned with still other ways of forging weapons from the insect's own life processes" (285).

Like many writers, Carson uses biblical references and cadences
to lend power and authority to important concepts. She compares
those in government and industry who deny the reported facts
about harm to wildlife from pesticides to "the priest and the Levite
in the biblical story" who "choose to pass by on the other side and
to see nothing" (86). In another ironic reversal, she writes: "By one
means or another, the new generations suffer for the poisoning of
their parents" (26).

Carson makes effective use of catalogs, in the manner of Walt
Whitman and Carl Sandburg. She gives remarkable internal variety
and balance to a list of warblers, interspersing descriptive phrases
between bare names: "the black-and-white, the yellow, the magno-
lia, and the Cape May; the ovenbird, whose call throbs in the May-
time woods; the Blackburnian, whose wings are touched with
flame; the chestnut-sided, the Canadian, and the black-throated
green" (111). The effect of this technique is to enhance the aesthetic value, and thus the importance, of what is threatened or lost.
Grammatical parallelism and repeating rhythms make sentences
surge toward a climax. This is the end of her description of the consequences of malfunction within the cell: "Then the muscle cannot
contract, nor can the impulse race along the nerve pathways. Then
the sperm cannot move to its destination; the fertilized egg cannot
carry to completion its complex divisions and elaborations" (203).

Carson often uses a figure of speech to help the reader understand a concept or process. A graphic simile demonstrates the process of energy production: "The transformation of matter into energy in the cell is an ever-flowing process, one of nature's cycles of renewal, like a wheel endlessly turning." Grains of fuel enter the wheel, go through successive changes, then finally emerge "stripped down," ready to combine with a new molecule and "start the cycle anew" (201). Later she compares a nonfunctioning cell to "a racing engine, generating heat but yielding no power" (203). For more complicated aspects of the process, the wheel image evolves into cell as "chemical factory" (201), with the mitochondria as "powerhouses" (202). Carson graphically warns us that "the crowbar to wreck the wheels of oxidation can be supplied by any of a number of chemicals commonly used as pesticides" (204). These graphic images let the nonscientific reader picture a complicated process by comparing it with something known—a return to the root purpose of figures of speech: to instruct in a clear and pleasing manner.

Carson reserves the greatest poetic effect for describing natural processes or beauty destroyed. She often combines sound devices—like alliteration—with pronounced mimetic rhythms. Explaining how chemicals spread through the earth's water supply, she writes of "spray that falls directly into streams or that drips down through the leafy canopy to the forest floor, there to become part of the slow movement of seeping moisture beginning its long journey to the sea" (40). In contrast, hard consonants—*ds*, *ps*, and *bs*—and frequent spondees (single-syllable stressed feet), as in "blizzards drive down," reflect the harsh beauty of life in the "land of the sage" (64). Interplay of related consonants, as well as straight alliteration, often connect phrases or paragraphs. In the "Fable," the "first *s*ettlers raised their houses, *s*ank their wells, and *b*uilt their *b*arns. . . . Then a *s*trange *b*light crept over the area" (2).

To broaden the perspective of the book, Carson introduces literary and cultural references. In rapid succession, she refers us to Greek mythology—Medea's robe, which brought its wearer violent death; Grimm fairy tales; and Charles Addams's macabre cartoons (32). These suggest that pesticide effects may surpass what we can imagine. "A house-that-Jack-built sequence, in which the large carnivores had eaten the smaller carnivores, that had eaten the herbivores, that had eaten the plankton, that had absorbed the poison from the water" (48) makes the passage of chemicals through the food chain into a grotesque joke on us.

The trials of a world filled with pesticides have an "Alice-in-Wonderland" quality for Carson. She quotes Lewis Carroll with angry humor:

> This system, however—deliberately poisoning our food, then policing the result [through government-set maximum permissible amounts, or "tolerances"]—is too reminiscent of Lewis Carroll's White Knight who thought of "a plan to dye one's whiskers green, and always use so large a fan that they could not be seen." (183–84)

Robert Frost's poem "The Road Not Taken" gives Carson a chapter title, "The Other Road," and a direct reference to introduce the final, forward-looking chapter of the book: "We stand now where two roads diverge." Carson elaborates this into an image of our traveling "with great speed" on "a smooth superhighway" with disaster at its end. She returns to Frost to introduce the alternative: "The other fork of the road—the one 'less traveled by'—offers our last, our only chance to reach a destination that assures the preservation of our earth" (277).

Reaching the Public

Philosophically, all of Carson's books and articles are of a piece. From her first book, the lovely narrative *Under the Sea Wind,* no matter what content, style, or rhetorical mode she employed, Carson transmitted the core of her ecological philosophy. "In each of my books," Carson told the Women's National Book Association, "I have tried to say that all the life of the planet is inter-related, that each species has its own ties to others, and that all are related to the earth. This is the theme of *The Sea Around Us* and the other sea books, and it is also the message of *Silent Spring*" (qtd. in Graham 53). What makes *Silent Spring* different is that the philosophy has become the overt message, that Carson's goal has become not education alone, as in the other books, but education in the service of persuasion: making the case for change.

Silent Spring works well because Carson has analyzed her potential audience, chosen literary and rhetorical techniques to reach them, and pragmatically organized the presentation of her ecological philosophy so that it will fit her readers' beliefs, concerns, and self-interest. Carson's letters demonstrate how carefully she orchestrated the process. In 1959, she explained to her editor, Paul Brooks, that it had always been her intention "to give principal emphasis to the menace to human health, even though setting this within the general framework of disturbances of the basic ecology of all living things." She expressed her confidence at being able to achieve "a synthesis of widely scattered facts" to build "a really damning case against the use of these chemicals as they are now inflicted upon us," paying particular attention to "the slow, cumulative and hard-to-identify long-term effects." She would also emphasize the positive approach, such as biological controls as alternatives to chemical sprays (qtd. in Brooks 243–44). She intersperses human-centered warnings and appeals among her calls for preservation of the greater environment, making sure the reader understands that "[m]an, however much he may like to pretend the contrary, is part of nature. Can he escape a pollution that is now so thoroughly distributed throughout our world?" (188). Carson assures us that we humans are as intimately interconnected with these problems as "the robin in Michigan or the salmon in the Miramichi" (189).

Contrasting Carson's approach in *Silent Spring* to that of the current philosophical movement of deep ecology demonstrates Carson's strategies for reaching and convincing the general public. The goal of deep ecology, as George Sessions explains in the preface to his collection of essays, is to redirect the "ecologically destructive path of modern industrial societies" through a shift in worldview from anthropocentrism, the religious and cultural belief that the earth and its creatures are there for the benefit of human beings, to ecocentrism, belief in the equal worth of all creatures. "The birth of the Deep Ecology movement paralleled the rise to public prominence of the science of ecology and the 'ecological perspective' as popularized by Aldo Leopold, Rachel Carson, and other ecologists." The deep ecologists, therefore, do not cite Carson as part of their "philosophical roots." She and Leopold, however, provided inspiration for the movement. Carson's concerns, according to Sessions, went beyond general questions about the threats of modern technology to human health "to encompass a respect and concern for the biological integrity of the Earth and all its species" and "posed a philosophical challenge to the *anthropocentrism* of Western culture" (ix–x). Arne Naess, founder and philosopher of deep ecology, maintains that supporters of the movement "ask deeper questions" and seek deeper social changes. "Like Rachel Carson, they tend to have firm convictions at a deep level" (205, 208). Deep ecologists, on the other hand, do not expect any implementation of their program in the near future, so they tend to reflect deep gloom in their writings. They are willing to countenance no interference with natural processes (Delany 28).

Here we must return to Carson's pragmatism. She had written pamphlets for the federal government promoting shared use of government lands for recreation and preservation and continued to reflect her belief in compromise and moderation in *Silent Spring*. She concentrated on the possible and sought a balanced approach to the "web of life." "Sometimes we have no choice but to disturb these relationships, but we should do so thoughtfully, with full awareness that what we do may have consequences remote in time

and place" (64). Although her philosophy suggests the central concept of deep ecology, that humans are not central to the universe of
creatures, she preserved the concept of stewardship: what we as humans brought about, we must correct as much as possible. We
might want to give up the dominance theory, but we must not discard responsibility along with it.

As argument directed to a specific audience, *Silent Spring* maintains a persuasive human-centered orientation: we must stop current practices not only for the sake of other creatures, or the earth
itself, but for our own sake. We are "more dependent on . . . wild
pollinators" (bees and other insects) than we usually realize (73).
Carson warns us that, although we rarely remember the "fact, [we]
could not exist without the plants that harness the sun's energy and
manufacture the basic foodstuffs [we] depend . . . upon for life"
(63). "Perhaps, even from our narrow standpoint of direct self-interest," she suggests satirically, "the relation [between weed and
soil] is a useful one" (78). Pragmatically appealing not only to our
biological and aesthetic but also to our economic self-interest, she
points out "the irony of this all-out chemical assault on roadsides
and utility rights-of-way":

> It is perpetuating the problem it seeks to correct, for as experi-
> ence has clearly shown, the blanket application of herbicides
> does not permanently control roadside "brush" and the spray-
> ing has to be repeated year after year. And as a further irony, we
> persist in doing this despite the fact that a perfectly sound
> method of *selective* spraying is known, which can achieve long-
> term vegetational control. (74)

She repeats this lesson throughout the book. Unlike the deep ecologists, and despite her underlying philosophy, Rachel Carson sought
neither to bring about a radical shift in worldview nor to begin a
new movement, although the environmental movement did flow

from reactions to her writing. She sought practical results based on bringing about limited and specific changes in public values.

Late in *Silent Spring,* there is a literary reference to the work of earlier writers. Like other references in the book, it broadens context and extends Carson's meaning. Discussing the relationship between animal experimentation and human experience, she uses the phrase "in men and mice," reversing the phrase John Steinbeck had earlier borrowed from Robert Burns. The lilt and alliteration of Carson's line—"in birds and bacteria, in men and mice" (207)—strengthens the poetic association. Burns, like Carson, often wrote of natural things. His poems, like her books, reflect profound respect for all life. When we look at the context of the phrase Carson has adapted, the final stanzas of "To a Mouse," we see how they bring further pungency to Carson's message:

But Mousie, thou art no thy lane [not alone],
In proving foresight may be vain:
The best-laid schemes o' mice an' men
      Gang aft a-gley,
An' lea'e us nought but grief an' pain,
      For promised joy.

Still thou art blest compared wi' me!
The present only toucheth thee:
But oh! I backward cast my e'e
      On prospects drear!
An' forward tho' I canna see,
      I guess an' fear!

The first stanza summarizes the lesson Carson has given us in *Silent Spring* with her history of the use of synthetic chemicals; the second captures the essence of her warning.

Note

1. Carson explained her approach in a letter (7 Jan. 1956) to Ruth Nanda Anshen (Brooks 331). Carson had contracted to write a book on evolution for a Harper series edited by Anshen (Lear 240).

Works Cited

Ashby, J. "Genetic Toxicity in Relation to Receptor-Mediated Carcinogenesis." *Mutation Research* 333 (Dec. 1995): 209–13.

Brooks, Paul. *The House of Life: Rachel Carson at Work.* Boston: Houghton, 1972.

Carson, Rachel. Papers. Yale Collection of American Literature. Beinecke Rare Book and Manuscript Library, Yale University.

———. *Silent Spring.* Boston: Houghton, 1962.

Coleman, Michael P., Jacques Esteve, Phillipe Damieki, Annie Arslan, and Helene Renard. "Trends in Cancer Incidence and Mortality." *IARC Scientific Publications* 121 (1993): 1–806.

Croy, R. G. "Role of Chemically Induced Cell Proliferation in Carcinogenesis and Its Use in Health Risk Assessment." *Environmental Health Perspectives* 101, Suppl. 5 (Dec. 1993): 289–302.

Delany, Paul. "D. H. Lawrence and Deep Ecology." *CEA Critic* 55 (Winter 1993): 27–41.

Goldberg, D. M., and E. P. Diamandis. "Models of Neoplasia and their Diagnostic Implications: A Historical Perspective." *Clinical Chemistry* 39 (Nov. 1993): 2360–74.

Graham, Frank, Jr. *Since Silent Spring.* Boston: Houghton, 1970.

Health Council of The Netherlands: Committee on the Evaluation of the Carcinogenicity of Chemical Substances. "Risk Assessment of Carcinogenic Chemicals in The Netherlands." *Regulatory Toxicology and Pharmacology* 19 (1994): 14–30.

Holleb, Arthur I., Diane J. Fink, and Gerald P. Murphy. *American Cancer Society Textbook of Clinical Oncology.* Atlanta: American Cancer Society, 1991.

Huff, James. "Issues and Controversies Surrounding Qualitative Strategies for Identifying and Forecasting Cancer Causing Agents in the Human Environment." *Pharmacology & Toxicology* 72, Suppl. 1 (1993): 12–27.

Lear, Linda. *Rachel Carson: Witness for Nature.* New York: Holt, 1997.

Miller, Elizabeth C., and James A. Miller. "Mechanisms of Chemical Carcinogenesis." *Cancer* 47 (1981): 1055–64.

Naess, Arne. "Deepness of Questions and the Deep Ecology Movement." Sessions 204–12.

Rowley, Janet, Professor of Medicine, University of Chicago. Personal interviews. 1 July 1981 and 23 May 1996.

Schottenfeld, David. "The Epidemiology of Cancer: An Overview." *Cancer* 47 (1981): 1095–1108.

Sessions, George, ed. *Deep Ecology for the Twenty-First Century*. Boston: Shambhala, 1995.

Rev. of *Silent Spring*, by Rachel Carson. *Economist* [London] 20 Oct. 1962: 248, 251.

Smith, Miles A. Rev. of *Silent Spring*, by Rachel Carson. *Press* [Bristol, CT] 6 Nov. 1962 (and in other newspapers). Clippings in Rachel Carson Papers, Beinecke Rare Book and Manuscript Library, Yale University.

Zimmerman, Hyman J. *Hepatotoxicity: The Adverse Effects of Drugs and Other Chemicals on the Liver.* New York: Appleton, 1978.

7

RANDY HARRIS

Other-Words in *Silent Spring*

Truth lives, in fact, for the most part on a credit system. Our thoughts and beliefs "pass," so long as nothing challenges them, just as bank-notes pass so long as nobody refuses them. But this all points to direct face-to-face verifications somewhere, without which the fabric of truth collapses like a financial system with no cash-basis whatever. You accept my verification of one thing, I yours of another. We trade on each other's truth. But beliefs verified concretely by somebody are the posts of the whole super-structure.

—William James, *Pragmatism*

Our practical everyday speech is full of the words of other people: we merge our voice completely with some of them, forgetting whose they are; others we take as authoritative, using them to support our own words; still others we people with aspirations of our own which are foreign or hostile to them.

—Mikhail Bakhtin, *Problems*

❧ Although we do with them what we will, every word we utter comes originally from someone else, from an Other. In the seeming paradox of Mikhail Bakhtin:

> words are not really ours; we only borrow them;
>
> but
>
> words are uniquely ours; they live in our peculiar heads.

The paradox resolves rather easily, of course: we are human. We borrow imperfectly. We borrow promiscuously. We borrow selfishly. We cram our borrowings up in private word hoards, where they nudge up against each other and fuse with subsequent borrowings of the "same" word. We make words ours, accidentally and deliberately. But echoes of Others nestle in each and every word we adopt,—in each and every belief we nurture, too, and (therefore) in each and every truth we hold, which come to us as configurations of words.

A small handful of linguistic devices have developed to preserve some level of Otherness in our speech—paraphrase, for instance, and quotation, and various conventions for flagging the word-hood of specific terms or phrases—they provide the solidest way to investigate the imprisoned choir of Others in our language. In this essay, I look at how Carson deploys these devices in *Silent Spring,* in the service of words, of beliefs, of truths.

Paraphrase and quotation differ largely in degree of replication and (when plagiarism is not at issue) in the type of flags used to signal the borrowing. They often cohabit quite comfortably. Here they are, living cheek by jowl in a typical example from *Silent Spring:*

> We are faced, according to Dr. Elton, "with a life-and-death need not just to find new technological means of suppressing this plant or that animal"; instead, we need the basic knowl-

edge of animal populations and their relations to their surroundings that will "promote an even balance and damp down
the explosive power of outbreaks and new invasions." (11)

Quotation marks flag the word-by-word nature of borrowing by
quote, and a name next to a verb of attribution gives us the source.
Only the name, the verb, and implicit reading conventions governed
by context convey the borrowing by paraphrase, which we understand is more conceptual than literal.

The natural cohabitation in this passage shows clearly how the
authority of the rhetor and that of her expert witnesses easily merge.
One minute Carson is talking about Charles Elton's position, the
next Elton is offering the position directly; and the first-person,
plural pronouns, especially in the paraphrased opening ("We are
faced"), leave little doubt that Carson endorses the position fully.
Here's another example of paraphrase and quotation rooming together in *Silent Spring,* though with a very different relation now of
rhetor to source:

> In a volume of *Proceedings* of one of the weed-control confer-
> ences that are now regular institutions, I once read an extraor-
> dinary statement of a weed-killer's philosophy. The author de-
> fended the killing of good plants "simply because they are in
> bad company." Those who complain about killing wildflowers
> along roadsides reminded him, he said, of anti-vivisectionists
> "to whom, if one were to judge by their actions, the life of a
> stray dog is more sacred than the lives of children." (71–72)

Carson's attitude toward the speaker is flatly contemptuous. His
words are characterized as "extraordinary"; he has no name, title,
or affiliation, only the epithet "weed-killer"; the inclusive *we* of the
previous passage is replaced by the definitionally exclusive (be-

cause singular) *I*, which stands in complete opposition to the reported words.

As one last example, consider the paraphrases and quotations
shaping this passage:

> A glance at the Letters-from-Readers column of newspapers al-
> most anywhere that spraying is being done makes clear the fact
> that citizens are not only becoming aroused and indignant but
> that often they show a keener understanding of the dangers and
> inconsistencies of spraying than do the officials who order it
> done. "I am dreading the days to come soon now when many
> beautiful birds will be dying in our back yard," wrote a Mil-
> waukee woman. . . .
>
> The idea that the elms, majestic shade trees though they
> are, are not "sacred cows" and do not justify an "open end"
> campaign of destruction against all other forms of life is ex-
> pressed in other letters. "I have always loved our elm trees
> which seemed like trademarks on our landscape," wrote an-
> other Wisconsin woman. "But there are many kinds of trees. . . .
> We must save our birds, too. Can anyone imagine anything so
> cheerless and dreary as a springtime without a robin's song?"
> (113–14)

The speakers are explicitly endorsed (and their ethos enhanced) by
putting their understanding in relief against the less-keen perceptions of government agents ("the officials"). Their words and emotions stand arm in arm with the author's.

These three passages illustrate the most immediate use to which
Carson puts Other-words in *Silent Spring*: to represent and re-pre-
sent three general constituencies whom—to profit from the con-
trastive clarity that caricature brings—I will label Good Guys (such
as Elton), Bad Guys (like the weed-killer), and Citizens (the Wis-

consin women). The Good Guys are scientists and related scholars who see the menace of insecticides clearly, and who offer alternative agricultural or forestry management methods. The Bad Guys are government policy makers, chemical-industry hacks, and unscrupulous entomologists more interested in lining their pockets than in preserving and studying insect life. The Citizens are homeowners, farmers, and other concerned nonspecialists who belong to nature societies, write letters to the paper, phone local agencies, and generally give vent to the bewilderment and outrage of people suffering the collateral damage inflicted by the Bad Guys.

But Other-words do not always come with explicit attribution. Drawing attention to the "word-hood" of words, as do the quotation marks of the previous clause, can also be a way of disengaging oneself from the literality of those words and therefore of flagging them as Other. On the literal skin here, I'm trying to deny Otherness. I'm saying, in effect, this is a unique and special Harris word. Revel in it. But of course the very act of congratulatory pointing makes inescapable that I borrowed *word* and *hood,* and even the hyphen, from the array of previous contexts I have found them in, and my only contribution is to string them together. The quotation marks allow me to engage in the childish academic sport of using a term while speaking beside my hand: "I know, I know: you won't find that term in the dictionary, and maybe it's a bit clumsy, but it works pretty nicely here; I'm clever." If you, in turn, were to write that "Harris is always prattling on about the 'word-hood' of words," beside your hand would run a commentary more on the order of "that word isn't in the dictionary; it's clumsy and affected; Harris thinks he's clever."

Carson's management of such devices goes in both these directions. Most often her use aligns with the latter, more complete, disengagement (as when she talks about the "safety" of certain insecticides, using the quotation marks simultaneously to identify the term as someone else's and to disavow its literal use). But she also deploys the mechanisms of disengagement in ways very similar to the former style (as when she writes of the "burning" consumption of energy in cells), which allow her to put words to figurative uses while sidestepping accusations of inaccuracy or poetical excess. And she uses them in a variety of other related ways,—for instance, commenting on the strength or requirements of her own argument, enhancing the status of a word by identifying it as the property of science (and boosting her own credibility by judiciously incorporating technical terms).

These are just the immediate purposes served by Other-words in *Silent Spring,* of course. From a broader vantage, seen in concert, they all work majestically to further truth, to advance belief, and—in the most prototypically rhetorical of their functions—to motivate action.

Paraphrase

> In Florida, fish from ponds in a treated area were found to contain residues of heptachlor and a derived chemical, heptachlor epoxide.
>
> —Rachel Carson

Like most science, journalism, scholarship, and any other fact-based discourse one might name, *Silent Spring* contains the words of Others obscured almost to the point of obliteration. Indeed, the obliteration of an Other, of an anybody, is the principal point of the empiricist repertoire—passives, clefts, all grammatical devices corrosive to agency—which gives fact construction such authority (Gilbert and Mulkay; Woolgar). A fact is hardly a fact if it can be said to belong to someone. When hundreds of fish showed up dead in the wake of the 1957 fire ant campaign, for instance, they "were found" by people other than Rachel Carson to contain pesticide residues (140). Never mind who scooped them out of the ponds, who took the sample, who tested their flesh, who read the instruments, who typed up the report that encoded and reified the findings,—never mind any of the whos down in Whoville: just give us the what. (In this case, Carson's immediate source appears to be a suitably anonymous circular from the Bureau of Sport Fisheries and Wildlife [U.S. Fish and Wildlife Service]; see her notes on p. 324.)

Take the following mininarrative:

A year-old child had been taken by his American parents to live in Venezuela. There were cockroaches in the house to which they moved, and after a few days a spray containing endrin was used. The baby and the small family dog were taken out of the house before the spraying was done about nine o'clock one morning. After the spraying the floors were washed. The baby and dog were returned to the house in midafternoon. An hour or so later the dog vomited, went into convulsions, and died. At 10 P.M. on the evening of the same day the baby also vomited, went into convulsions, and lost consciousness. After that fateful contact with endrin, this normal, healthy child became little more than a vegetable—unable to see or hear, subject to frequent muscle spasms, apparently completely cut off from contact with his surroundings. Several months of treatment in a New York hospital failed to change his condition or bring hope of change. "It is extremely doubtful," reported the attending physicians, "that any useful degree of recovery will occur." (27)

Silent Spring is full to brimming with such stories, most without explicit attribution, and it is easy enough to see that their words had an active life before they made it through Carson to the pages of the book. In this case, they come most directly from "Poisoning by Endrin," a brief by Harold Jacobinzer and Harry W. Raybin, in a series on accidental chemical poisonings in the *New York State Journal of Medicine.* Jacobinzer and Raybin, in turn, are recounting a case originally reported by Dr. Edmund N. Joiner III, then Chief of Pediatrics at Roosevelt Hospital, in New York City. It's not clear whether (but not particularly likely that) Joiner was the attending physician, but certainly he had others—technicians and nurses as well as doctors—provide him with some of the clinical results, information about symptoms, details of family history, circumstances of exposure, and so on. Since the poisoning took place in Venezuela, some of the words must have come from there as well. Sometimes these are identified *as* words and linked to a generic speaker ("[A] veterinarian stated that he thought the [dog's] death was due to ingestion or absorption of the Endrin" [Jacobinzer and Raybin 2018]). Sometimes both their linguistic and personal origins are obscured, though it remains clear that the words come from others ("[W]hen first seen, [the baby] was having generalized tonic and clonic convulsions" [2018]). Several other clumps of words have obvious sources beyond Jacobinzer, Raybin, or Joiner ("It is interesting to note that according to the father [the victim's] pattern of likes and dislikes of food were exactly the same as before his illness" [2021]). Indeed, even the poisoned baby has words that participate in the report:

> At the time of his illness he was an alert, well-developed, well-integrated baby who was crawling and beginning to take a few steps alone. He held his bottle in his hands, recognized his parents and siblings, vocalized well, and was beginning to say a few simple words, like "da da" and "mama." (2017)

Carson's treatment of this narrative is not especially remarkable, except perhaps for the skill with which the paraphrase is effected. There is a direct quotation,[1] and a few of the more formulaic phrases and terms, like "a spray containing Endrin" and "after the spraying the floors were washed," come almost verbatim from the original brief (see Jacobinzer and Raybin 2018). Otherwise, there is no indication that the story owes anything to the words of Others.

Some shaping toward stronger claims, added drama, and pathos is evident from Carson's rephrasing. Jacobinzer and Raybin's "gradual, definite but slight improvement" (2021), for instance, becomes her "[hospital care] failed to change his condition or bring hope of change." Their "exposure" becomes her "fateful contact." Their preferred terms for the victim are "infant" and "child" and "patient"; Carson's is "baby." But increasing the strength of claims and instilling drama are extremely common when material moves into public discourse from the more tentative and qualified world of scientific literature (Fahnestock; Myers), and focusing emotion is one of the most recurrent and effective elements of Carson's style. (In fact, while Carson does crystallize the pathos into a few words like "baby" and "vegetable," the long catalog of the child's conditions and disorders that Jacobinzer and Raybin give in excruciatingly clinical detail is far more emotionally unsettling. Carson manages the pathos; she does not exploit it.)

Carson tells other types of stories, many of them so conventionalized that they no longer look like stories. But they follow much the same pattern, except that the sources are even more remote. All of the insecticides she discusses, for instance, come in for narrative treatment. Indeed, the story of the child is just the last episode in the Tale of Endrin:

> Endrin is the most toxic of all the chlorinated hydrocarbons. Although chemically rather closely related to dieldrin, a little twist in its molecular structure makes it 5 times as poisonous. It makes the progenitor of all this group of insecticides, DDT, seem by comparison almost harmless. It is 15 times as poison-

ous as DDT to mammals, 30 times as poisonous to fish, and about 300 times as poisonous to some birds.

In the decade of its use, endrin has killed enormous numbers of fish, has fatally poisoned cattle that have wandered into sprayed orchards, has poisoned wells, and has drawn a sharp warning from at least one state health department that its careless use is endangering human lives. (26–27)

We get its endowments, its genealogy, its accomplishments—just the sort of features we get for Achilles or Beowulf—before we move to the havoc it wreaks on the child in Venezuela. The words of the Tale of Endrin all have sources prior to Carson, although there is no attribution for this material.[2] Nor should there be. It is a collection of facts, but their collectivity, and their facticity, should not obscure the way in which the words are petrified. Indeed, their status as facts is in large part determined by the boilerplate phrasing that linguifies them. We would not be surprised to find that a phrase like "the most toxic of all the chlorinated hydrocarbons" came verbatim from something Carson had read, perhaps long before (re-)committing this configuration of words to the pages of *Silent Spring*; or that a clause like "endrin has killed enormous numbers of fish" is effectively a précis of some technical report she had consulted. (Let me be clear: I am not accusing Carson of plagiarism here, even "unconscious plagiarism," and still less of "unoriginality"; I am talking about general processes in the cycles of knowledge and language represented by this sort of paraphrasing and narrative structuring.)

Stories and events and facts bulge from every page of *Silent Spring*, sometimes with clear sources in the text but most often without, and they all have essentially the same character. Carson takes in, repackages, and reframes the words of Others in ways that build her case. In Bakhtin's terms, Carson-as-narrator in *Silent Spring* is the "ultimate semantic authority," and her use of material from elsewhere is simply part of the scaffolding required by the design of her argument (*Problems* 155).[3]

Quotation

> . . . one point of view opposed to another, one evaluation
> opposed to another, one accent opposed to another . . .
> —Mikhail Bakhtin, *Dialogic Imagination*

Other-words, in Bakhtin's handling, come in two basic versions.[4] Sometimes they are oblivious to their own incorporation, "like a man who goes about his business unaware he is being watched" (Voloshinov [Bakhtin] 197). Sometimes they are "penetrated" by the author, infiltrated, put to ends different from what the original speaker intended, possibly opposite ends, and Bakhtin gets a trifle sinister in his language about this style of incorporation:

> [T]he other speech act is completely passive in the hands of the author who avails himself of it. He, so to speak, takes someone else's speech act, which is defenseless and submissive, and implants his own intentions in it, making it serve his own aims. (Voloshinov [Bakhtin] 198)

Each mode of incorporation, in turn, has several subdivisions, depending on which Bakhtin one consults, but for our purposes only these two extremes are relevant. At one end is the type of incorporation that is "unmediated" and "intentional," in which the reported speech "constitutes the ultimate conceptual authority within the given context" (197). At the other end is the type of incorporation that thumbs its nose at the reported speech; the author's voice, "having lodged itself in the other speech, clashes antagonistically with the original, host voice and forces it to serve directly opposite aims" (198). Carson uses the first mode for the words of Good

Guys (experts she endorses) and for the words of Citizens; she uses the second mode, which Bakhtin aligns with parody, for the words of Bad Guys (chemical-industry spokespeople, some governmental agencies, and anyone who holds pro-insecticide positions).

Carson individualizes the Good Guys, particularizing them and framing their positions very favorably. A typical example is William O. Douglas, who always shows up wearing his title "Justice" (67, 68, 72, 159), and whom Carson describes as a "humane and perceptive jurist" (72). C. J. Briejèr is regularly Dr. Briejèr (78, 273, 275 [twice]), is "director of the Plant Protection Service" in Holland (xiii, 275), and is introduced as "a Dutch scientist of rare understanding" (78). Robert Cushman Murphy is "Curator Emeritus of Birds at the American Museum of Natural History" (103), "one of the world's leading ornithologists" (103), and "the world famous ornithologist" (159). W. C. Hueper is "Dr." (18, 50, 221, 222, 225, 235, 239 [twice], 243 [twice], 241 [thrice], 242; also "M.D.," xiii). He is "of the National Cancer Institute" (xiii, 18, 50, 221). He is "an authority on environmental cancer" (18), indeed "a foremost authority on environmental cancer" (221). He has the "thoughtfulness of one who has pondered long" (240). He "has a lifetime of research and experience behind his judgment" (240). He is the author of "a classic monograph on the subject [of environmental cancer]" (222). Many of "the most eminent men in cancer research" share his beliefs (242). His "years of distinguished work in cancer make his opinion one to respect" (240).[5]

When the words of the Good Guys come without the company of a specific name, it is almost always in the tones of detached authority characteristic of science. They come from such sources as "Scientists of the Dutch Plant Protection Service" (78) and "the biologists" (134), and "a scientific team from the United States Public Health Service" (178) and "water pollution experts" (238).

In sharp contrast, the Bad Guys are almost always nameless, anonymous, without titles or accomplishments. Their affiliations loom large—to government agencies or chemical companies—but often we get nothing else of them, only their words. By establishing the eminence of her expert witnesses, the Good Guys, Carson effectively puts them out of her own reach. She does not have the authority to question their findings, their positions, or their arguments. But then, she doesn't want to question their words, just to pass them along to her readers; indeed, she strengthens and enhances their authority by her reverential handling of their words. Leaving the Bad Guys as faceless bureaucrats and salesmen gives her the room to penetrate their utterances,—undermining, rejecting, and ridiculing them. And, of course, readers are free to fill in the attributes or faces of every inert government functionary who has ever obstructed them, or of every glib, oily salesman who has sold them a lemon.

But Carson departs from this fill-in-the-blank strategy in two marked ways. On the rare occasions when these people have labels, they are virtually always mocking epithets: we get the words of "a weed killer" (72), the beliefs of "the man with the spray gun" (86), the judgments of "the control men" (86), the decisions of "the authoritarian" (127). (Sometimes their own labels, like "agricultural engineer," are turned against them; see below.) And anticipating the objection that many respected scientists promote the use of insecticide (it isn't clear if any of them are among the faceless voices she quotes), she actively erodes their ethos. One passage that is particularly important in this regard concerns the vast sums of money that pour into universities from the major chemical companies. Carson says this situation

> explains the otherwise mystifying fact that certain outstanding entomologists are among the leading advocates of chemical control. Inquiry into the background of some of these men reveals that their entire research program is supported by the chemical industry. Their professional prestige, sometimes their

very jobs depend on the perpetuation of chemical methods. (259)

"Can we then expect them to bite the hand that literally feeds them?" she adds (leaving the enthymematic "no, of course not," to her readers). And, more particularly, she asks about their (lack of) credibility: "[K]nowing their bias, how much credence can we give to their protests that insecticides are harmless?" (259).[6] While she is explicitly on the subject of ethos, she quickly makes note of other scientists, the ones who disagree with chemical control—"those few entomologists who have not lost sight of the fact that they are neither chemists nor engineers, but biologists" (259)—backhandedly suggesting that the Bad Guys don't even know the appropriate labels for their own work.

The Citizenry is Carson's largest constituency. Like the Bad Guys, but for inverse reasons, the Citizens too are mostly nameless and faceless. Their authority, and it is considerable in the text, comes from their numbers, from the passion they bring to the argument, from the weight of their eyewitness accounts, and from their closer alignment with the majority of Carson's readers. (The great majority of these Citizens are women.)

The first Citizens (a "New England woman" and "a conservationist") show up as clear representatives of ground-swelling outrage. The very first in the book, writing "angrily" to a newspaper, speaks as part of a "steadily growing chorus of outraged protest about the disfigurement of once beautiful roadsides by chemical sprays" (69). The second, "angry at the desecration of the Maine roadsides" (70), is a more particularized member of that chorus, one of a group of conservationists who visited a recently sprayed and rapidly browning Maine island (69). The next group we hear from as a flurry of excited calls to the Detroit Audubon Society, after a chemical campaign against the Japanese beetle killed large numbers of birds and incapacitated pets (90); local sportsmen in Illinois quickly join their voices to the eyewitness accounts of dead and dying birds (and rabbits and muskrats and opossums and fish).

But it is with the pivotal chapter, "And No Birds Sing," that the Citizen chorus really raises high the roof beams. The title of that chapter recalls the book's chilling epigram from Keats, reaches into the opening "Fable" to remind readers of the central loss there ("The birds, for example—where had they gone? Many people spoke of them, puzzled and disturbed" [2]), and evokes the ominous ruling metaphor of the entire book, a silent spring: the chorus serves that title and its resonances well. The chapter opens with "a housewife," writing "in despair" of the consequences of several years of DDT spraying for the bird population, adding the voices of her children to the chorus:

> [T]he town is almost devoid of robins and starlings; chickadees have not been on my shelf for two years, and this year the cardinals are gone too; the nesting population in the neighborhood seems to consist of one dove pair and perhaps one catbird family.
>
> It is hard to explain to the children that the birds have been killed off, when they have learned in school that a Federal law protects the birds from killing or capture. "Will they ever come back?" they ask, and I do not have the answer. The elms [which the spraying program was meant to protect] are still dying, and so are the birds. *Is* anything being done? *Can* anything be done? Can *I* do anything? (103)

An "Alabama woman" writes in *Audubon Field Notes* that within months of the fire ant campaign in her area "there was not a sound of the song of a bird. It was eerie, terrifying" (104). "One woman" calls the Cranbrook Institute of Science to describe a dozen dead robins on her lawn (109). A "Milwaukee woman" writes of "the pitiful, heartbreaking experience" of finding beautiful birds dying in her backyard (113–14). A "Wisconsin woman" asks if anyone can "imagine anything so cheerless and dreary as a springtime without a robin's song" (114). In England, "a landowner" complains that "the destruction of wildlife is quite pitiful" (123); a game-keeper writes that finding hundreds of dead birds is "a distressing experience for me. It is bad to see pairs of Partridges that have died together" (123); and witnesses testify to avian death and disappearance to the House of Commons (124).

The Good Guys and the Bad Guys, as Good Guys and Bad Guys always do, determine the drama. They poison and protest and argue, and poison and protest and argue some more. Citizens are largely spectators. They're vocal spectators, writing letters to the paper and phoning local conservation societies about what they've seen and how they feel, but they are spectators all the same, which makes them the closest constituency in the book to Carson's implied audience. She has no illusion about reaching the Bad Guys directly, and the Good Guys are part of what Wayne Booth calls the "Community of the Blessed." They already believe in the case she is arguing. But the everyday, purchasing, voting, song-bird-appreci-ating Citizen can not only be brought in to believe but might be moved to action.[7] By giving so many Citizens a voice in *Silent Spring,* Carson is also giving voice to her readers, engaging them in the book and in the argument.

Further than that, she provides them with a template for future action. The nonspecialists she quotes are model citizens in the literal sense that they provide models of conduct, even models of writing, for Carson's readers to emulate in the pursuit of legislation governing responsible pesticide use.

Although she was herself a scientist, and certainly qualified as an expert on insecticides by the time *Silent Spring* was published, it is among the Citizenry (therefore, the audience), that Carson most frequently places herself, especially when she uses first-person pronouns to step out from behind the persona of detached authority. "We are told," she says at one point, "that inoculation with milky spore disease is 'too expensive'"; and "We are told also that milky spore disease cannot be used on the periphery of the beetle's range" (98)—when the source of information in both cases is a letter Carson received (presumably in response to a query) from the Fairfax Biological Lab (317). The *we* is Carson, representing the rest of us. Carson also joins in the eyewitness accounts of New England roadside devastation (71) and places herself squarely among the people trying to sort through the conflicting evidence of experts and officials:

> The citizen who wishes to make a fair judgment of the question of wildlife loss is today confronted with a dilemma. On the one hand conservationists and many wildlife biologists assert that the losses have been severe and in some cases even catastrophic. On the other hand the control agencies tend to deny flatly and categorically that such losses have occurred, or that they are of any importance if they have. Which view are we to accept? (86)

The *we* of that last question is significant for the primary role Carson assumes in the book—just another Citizen helping others find the truth by mediating between experts—and the answer she gives that question is even more significant. (See 178–84 for a more extended example of this mediation.) "The credibility of the witness is of first importance," she writes (86; agreeing with Aristotle 1356a). "The professional wildlife biologist" we can trust because he "is certainly best qualified to discover and interpret wildlife loss," and he is "unbiased in favor of chemicals." The "bird watcher," too, as well as "the suburbanite who derives joy from birds in his garden, the hunter, the fisherman," and "the explorer of wild regions," have "valid point[s] of view." In short, we can trust experts of the right sort, Good Guys, and we can trust our fellow Citizens. The "entomologist," on the other hand, is "not so qualified" as the biologist and "is not psychologically disposed to look for undesirable side effects of his control program"; still less trustworthy are "the chemical manufacturers" (86).

But Carson also adopts the voice of the Good Guy expert quite consistently, sometimes even stepping out of the mass of Citizens to give her own expert testimony. Usually, this adoption is subtle and well behind the scenes, by way of the objective tone with which she paraphrases and frames the case of experts against widespread insecticide use. On a few occasions, though, Carson puts words directly in the mouths of her readers and then answers them back authoritatively. "To the question 'But doesn't the government protect us from such things?'" which Carson supplies for her readers, she answers, "Only to a limited extent" (181). "'But,' someone will object," she notes, "'I have used dieldrin sprays on the lawn many times but I have never had convulsions like the World Health Organization spraymen—so it hasn't harmed me.'" She quickly adds, "It is not that simple"; she then presents a dramatic case of long-term exposure (190).[8]

Words Being Words

> They should not be called "insecticides" but "biocides."
> —Rachel Carson

In the last paragraphs of *Silent Spring,* Carson tells us that the Bad Guys have committed worse sins than the perversions of avarice. They are dangerously prideful, lacking "humility before the vast forces with which they tamper" (297). Capturing this dangerous pride, in fact, is a phrase that sums up the appalling philosophy that fuels insecticide use:

> The "control of nature" is a phrase conceived in arrogance, born of the Neanderthal age of biology and philosophy, when it was supposed that nature exists for the convenience of man. The concepts and practices of applied entomology for the most part date from that Stone Age of science. It is our alarming misfortune that so primitive a science has armed itself with the most modern and terrible weapons, and that in turning them against the insects it has also turned them against the earth. (297)

This passage is worth quoting for the way it ensures that readers are left with an inescapably poor image of the Bad Guys, as blunderingly arrogant, slope-browed agents of destruction. But it also introduces a use of Other-words that falls outside of paraphrase or direct quotation. The phrase "control of nature" isn't cited directly from anyone here. It is just given as a common and traditional label for the enterprise in which the Bad Guys are engaged. Quotation marks here don't signal that the phrase has come from some particular mouth, pen, or typewriter. They flag it as being, in the parlance of language philosophers, "mentioned." If I tell you that this page is made of paper, I am *using* that noun; if I tell you that "paper" has five letters or two syllables or follows the preposition "of" in the last clause, I am *mentioning* it—talking *about* it rather than *with* it, pointing at it rather than putting it to work describing the world. (This is true, of course, whether or not I observe the convention of putting it in quotes or if I use some other convention, like italics.) In the passage above, Carson is mentioning "control of nature" rather than using it to directly talk about the practices of the Bad Guys (though she quickly comments on those practices with her discussion of why that phrase was "conceived in arrogance").

Mentioning words or phrases is the prototypical use of quotation marks when they aren't actually flanking words being attributed directly to someone else, and Carson puts them to this use frequently. For instance, she regularly places technical terms, like "dinitro" (36) and "potentiated" (31) and "sterilants" (56), in quotation marks because she is introducing the words to her readers, going slightly above the diction of her audience.[9] For symmetrically opposite reasons, she puts a number of colloquial expressions, like "silo deaths" (78) and "milky disease" (97) and "sacred cows" (114), in quotation marks, perhaps to acknowledge she is momentarily dropping below the accepted diction for a popular science book (of the early sixties).[10] In both cases, however, she is not just mentioning the words. She is also using them to refer to phenomena and, in some cases, descriptively enliven those phenomena. But she is talking about the word-hood of the words, about their status as pieces of language, as well as using them. On one occasion, she brings out quotation marks to signal a grossly inadequate term and the dangerous implication that follows from not having access to more adequate terms: "Often [these chemicals] cannot even be identified. In rivers, a really incredible variety of pollutants combine to produce deposits that the sanitary engineers can only despairingly refer to as 'gunk'" (40).

The common term for this device is "scare quotes," and as this term suggests, extra-quotational quotation marks don't always adhere to the prototype of mentioning.[11] Carson uses scare quotes very effectively, in a way that fits their popular label like a glove: undermining the claims of government and chemical-industry spokespeople about the value and safety of insecticides, ironizing the claims to scare her readers away from the uncritical acceptance of government and corporate discourse.

Drawing attention to the word-hood of a word sometimes serves little more purpose than saying "something is up with this word; I'm not using it in a straightforward, literal way." For instance, Carson says that "after the disastrous 1957 [gypsy moth] spraying the program was abruptly and drastically curtailed, with vague statements [from the Department of Agriculture] about 'evaluating' previous work" (161). She might have left the quotation marks off *evaluating*, and the remark would be denotatively the same: the statements were vague and they concerned evaluating previous work. Bagging *evaluating* in quotation marks, though, says that this word is suspect. Maybe they were lying. Maybe they were covering ineptitude and irresponsibility. Maybe they were retroactively redefining the program as a pilot rather than a campaign. But whatever is going on, we shouldn't take *evaluating* at face value when we hear it from Bad Guys; in fact, we shouldn't take very many of their words at face value.

The most extended example of scare quotes to undermine the credibility of the Bad Guys in *Silent Spring* is the recurrent flagging of words like *safe* and *harmless* in connection with control chemicals:

The alleged "safety" of malathion. . . . Malathion is "safe" only . . . (31)

. . . careless use in the belief that they are "safe" can have disastrous results . . . (35)

. . . they had acted on advice that 1 pound of DDT to the acre was "safe" . . . (136; also 180, 182, 224, 232, 237 [twice], 238, 239)

. . . what dangerous substances may be born of parent chemicals labeled "harmless"? (44)

Too many of them are to be found defending herbicides as "harmless" . . . (74)

. . . showers of "harmless" poison descending on people . . . (90; also 107 [twice], 191)

Occasionally these terms show up in bona fide quotations:

> An official of the Federal Aviation Agency was later quoted in
> the local press to the effect that "this is a safe operation" and a
> representative of the Detroit Department of Parks and Recrea-
> tion added his assurance that "the dust is harmless to humans
> and will not hurt plants or pets." (89)

But even without these attributions, it is clear that the terms come
from officials and representatives and spokespeople for government
agencies and chemical manufacturer,—from the Bad Guys. And it
is equally clear, through the scare quotes alone (though there is, of
course, plenty of other evidence to that end in the book), that these
people are at best deluded, at worst lying, that the chemicals are the
opposite of safe and harmless. Carson draws attention to the word-
hood of these terms to show how they are being misused. In turn,
this implicit commentary undermines the credibility of those voices.

Closely related to this outright reversal of meaning is Carson's
treatment of words like "pests," "brush," and "weeds." "Since the
mid-1940s," she tells us, "over 200 basic chemicals have been created for use in killing insects, weeds, rodents, and other organisms
described in the modern vernacular as 'pests'" (7). Here the label is
flagged not because of the source or because of falseness, but because of the perspective it represents, a perspective Carson rejects.
An organism is only a "pest" from a highly anthropocentric view;
from a more embracing position, it is just another node in the web
of life, no different from any other natural organism (and, in most
cases, a good deal less harmful to the environment than the extra-
evolutionary domestic organisms created by humans).[12] Similarly,
she identifies the use of "weeds" as the product of a narrow perspective (which she links more directly to the Bad Guys). After a
lyrical paragraph on sweet fern, juniper, clover, purple vetch, and
other wild plant life, she says "such plants are 'weeds' only to those

who make a business of selling and applying chemicals" (71; also 72, 73, 79; 68 for "weed out"; see 69, 74, for "brush"; 68 for "brush control"; 68 for "improvement").

The common term for this phenomenon, at least when it is used in speech with a peculiar two-handed bi-digit–wiggling gesture, is "air quotes." The device says "other people use this term, but I can't really bring myself to use it directly, and you should regard it as fishy." In stodgier discourse, such uses are often proceeded by terms like "so-called."

Carson also uses this technique in her most explicit critique of the Bad Guys' language. For instance, in one double-purpose passage, she first snickers at the label they choose for themselves and then condemns the ignorance of the terminology with which they describe their work: "The 'agricultural engineers' speak blithely of 'chemical plowing'" (69). Other blithe euphemisms Carson ridicules via quotation marks include "improvement" (chemically soaking forests [68; see also 50]), "sickness" ([125] a condition better described as "poisoning"), and "tolerances" for the "maximum permissible limits of [food] contamination [by insecticides and/or herbicides]" (181). Additionally, Carson singles out "eradication" for special treatment, in connection with the U.S. Department of Agriculture's disastrous gypsy moth and fire ant campaigns of the late 1950s (157–72). Bringing the word up repeatedly in chronicling those campaigns, almost always bagged in quotation marks, serves to show simultaneously the hubris of the department and its extremely dangerous blundering.

Neither the words of the Good Guys nor those of the Citizenry come in for much extra-quotational quotational treatment. Among the Good Guys, Carson puts a few technical terms in quotation marks, like Hueper's "chemical carcinogen" (225), and she lets Dr. L. D. Newsom, director of entomology research at the Louisiana State University Agricultural Experiment Station, share in her condemnation of the Bad-Guy term "eradication" (172). She also borrows the word "impossible" from a quotation of Briejèr (245–46) simultaneously to illustrate his astuteness and to warn about the shortsightedness of broad chemical assaults on insects. And among the Citizenry, she flags several hopeful words about cancer research to suggest that the concepts they designate are mythical (and, therefore, that prevention is paramount): "magic bullets" (240), "wonder drugs" (240), "cure" (241 [twice]), and "breakthrough" (241). But that's it.

She does, however, use quotation marks frequently to comment on her own argument and language use. In particular, she twice flags phrases like "it is admittedly difficult . . . to 'prove' that cause A produces effect B" (192; see also 220), suggesting that demands for the corresponding level of rigor are unreasonable. And she regularly flags her terminology whenever she fears it might be getting a little too poetical. Her purpose is a desperately serious one, and she worried about vulnerability to the charge of straying from the facts, of getting emotional, of being too literary in her treatment of matters scientific. So she was particularly careful to keep the reins on her terminology. In the chapter on cell biology and genetics, for instance, she regularly flanks terms like "burning" (200) and "powerhouses" (202) and "racing engine" (203), and even "instructions" (214), with quotation marks to signal that she is aware that she is talking in images rather than concrete specifics. When she stretches her regular radiation/insecticide analogy a little too far, she puts "fallout" in quotation marks (229). When she uses "tough" and "toughness" of insects (273), or "mistakes" of cellular processes (284), they get flanked to signal that the anthropomorphism is deliberate. When she evokes Frost's famous poem, the phrase "less traveled by" is marked (277). When she waxes a bit eloquent with a phrase like "the treadmill of chemical control," out come the quotation marks (279).[13]

Conclusion

What then can be a "safe dose" of DDT?
—Rachel Carson

If someone asks you what a "safe dose" of DDT is, as Carson just has, you know right away that the words in quotation marks are not hers. She is disavowing them. She has brought them in from elsewhere and bagged them in pairs of raised commas, one pair inverted, so you will know that someone else—someone without enough knowledge, or someone with duplicitous intentions—has used them in a way that is misleading and dangerous. The words are not hers, but you have no trouble recognizing that she has a belief about them or that the belief concerns their veracity. The quotation marks deliver a truth judgment, a negative truth judgment, on the material they flank. There *is* no safe dose of DDT, Carson says with her quotation marks, and you shouldn't believe anyone who tells you there is. Compare this question to the all but identical, "What, then, can be a safe dose of DDT"? There is some distancing in the use of *then* and *can be*, but there is no immediate implication that the underlying presupposition is false; indeed, the default implication is that there is a safe dose of DDT, which we just need to find.

Bakhtin (though his immediate target is elsewhere) hits precisely on the effect with which Carson punctuationally skewers government and corporate spokespeople when she borrows their words: "The incorporated languages and socio-ideological belief systems . . . are unmasked and destroyed as something false, hypocritical, greedy, limited, narrowly rationalistic, inadequate to reality" ("Discourse" 311–12). "False," "hypocritical," "greedy," "limited," "narrowly rationalistic," "inadequate to reality": there may not be a better string of adjectives to describe the official discourse around insecticide use, as it is revealed in *Silent Spring*.

Carson penetrates and explodes the discourse of insecticide proponents. Other discourse she swallows whole:

> "We all live under the haunting fear that something may corrupt the environment to the point where man joins the dinosaurs as an obsolete form of life," says Dr. David Price of the United States Public Health Service. "And what makes these thoughts all the more disturbing is the knowledge that our fate could perhaps be sealed twenty or more years before the development of symptoms." (188)

Carson, it is clear, endorses these borrowed words. The endorsement is not just in the overwhelmingly positive context Carson provides, not just in the ethos-enhancing use of the speaker's title and affiliation, but in the very fact that there is a (nonpenetrated) quotation at all. Price's words are elevated by the borrowing alone, declared to be more important than anything Carson might have said in their stead, because of their eloquence, because of their authority, because of their urgency, or just because she wants us to know she is not isolated in her concerns. Here, we are at the opposite end of the epistemic scale, where the truth judgment offered on the flagged words is ringingly positive.

Other words yet she reconfigures. The ballad of That for Which There Is No Safe Dose, for instance, begins

> DDT (short for dichloro-diphenyl-trichloro-ethane) was first synthesized by a German chemist in 1874, but its properties as an insecticide were not discovered until 1939. Almost immediately DDT was hailed as a means of stamping out insect-borne disease and winning the farmers' war against crop destroyers overnight. The discoverer, Paul Müller of Switzerland, won the Nobel Prize. (20)

Carson doesn't say where this information comes from, and we wouldn't expect her to. It is "common knowledge" (albeit not commonly known) of the sort available in any encyclopedia. The phrase "was first synthesized by a German chemist in 1874" or "short for dichloro-diphenyl-trichloro-ethane" might easily have come word for word from a book Carson had read; certainly she must have seen at least very similar configurations.

Here, we go right off the (positive) end of the epistemometer. By telling the story of DDT in "her own words," Carson submerges the notion that truth is even an issue. She gives that story the imprimatur of her voice, the author's primary voice. The Other-words partake of her "ultimate conceptual authority" in an even more compelling way than the sort of (quotational) borrowing Bakhtin brands with that phrase. There is not even the potential mitigation of "according to Dr. Elton . . . " or "says Dr. David Price" or "The *Encyclopaedia Britannica* describes DDT as . . . " There is not even the potential mitigation of a loaner behind the borrowed words. Carson gives them to us directly and takes complete responsibility for them.

Carson's responsibility for the words, the beliefs, and the truths of *Silent Spring* is the wellspring of its power. Her ability to sift through an enormous body of Other-words and tell us which ones to believe, which ones are true, which ones we should act upon; her ability to bring us the words of experts, to puncture the self-serving and dangerous words of an industry, to unify the outraged words of a citizen chorus, to make us think about the authority that wordhood confers; her ability to leave us all these words in configurations that become our own, makes *Silent Spring* a rhetorical masterpiece. There are many reasons for using the words of others, but the most compelling reason in this context—one that is especially important in texts about science, and completely essential in polemical texts about science, of which *Silent Spring* is the premier example—is to loan out beliefs so strong that they acquire the status of truths and become knowledge. Carson wants her readers not just to *feel* that insecticides are dangerous, not even just to *believe* that insecticides are dangerous, but to *know* that insecticides are dangerous. And to *do* something about it.

Notes

I would like to thank Jimmie Killingsworth, Jackie Palmer, Craig Waddell, and three (I regret to say) anonymous reviewers for very helpful comments on earlier versions of this essay. I would also like to thank Catherine Schryer, Bill Schipper, and the "Beer and Baxtin" discussion group for stimulating, polyphonic chats; Sonja Sen for her research on rhetoric in *Silent Spring*; and Greg Cento for (indirectly) suggesting the thesis of this essay.

1. Actually, Carson gets the quotation a bit wrong, changing "it seems extremely doubtful" to "it is extremely doubtful," and giving the source as "the attending physicians" (when the quotation is of Jacobinzer, an Assistant Commissioner for the New York City Department of Health, and Raybin, Technical Director of that department's Poison Control Center), though nothing of substance is affected by the slips.

2. Aside from the Jacobinzer and Raybin article (which does not discuss such details), an unavailable mimeograph is the only source cited among the reference notes for p. 26 that includes "endrin" in the title.

3. More fully, Bakhtin talks about "the direct speech of the author" from the perspective of the word(s) she assumes in her own voice:

> The direct object-oriented word recognizes only itself and its object, to which it seeks to conform in the highest degree possible. If in the process it imitates or takes a lesson from someone else nothing is changed in the slightest: it is like the scaffolding which is not part of the architect's plan, although it is indispensable and is taken into consideration by the builder. The fact that another person's word is imitated and that the words of other persons, clearly recognizable to the literary historian and to every competent reader, are present is not of concern to the word itself. (*Problems* 155)

4. I am talking here of the types of reported speech he calls "single-voiced" and "double-voiced" (also, in less-revealing terminology, "linear" and

"pictorial" [Voloshinov (Bakhtin) 199–223]). This is as good a place as any to acknowledge that the work of Bakhtin is very problematic. He revisits the same topics through different lenses, often in quite different contexts, on a regular basis, and he pays little obeisance to consistency. My "Bakhtin" in this paper, as all "Bakhtins" must be, in all discussions, is a fictional composite, assembled from remarks in scattered places and circumstances and somewhat overlaid with my own notions of coherence and systematicity. With a little inspection, of course, all discursive representations of others, perhaps even all discursive representations of ourselves, even in such determinedly factual genres as (auto)biography, prove to be fictional composites. In fact, that's one of the implications of Bakhtin's work. It's just that much less inspection is required to find the selectivity in anyone using Bakhtin's words because he often says something significantly different elsewhere. One specific remark of his that runs counter to my analysis, though, is worth mentioning. Although Bakhtin is used increasingly by rhetoricians, most of his commentary is directed at literature, and the great majority of his examples come from literature; rhetoric, he is less interested in, and less perceptive about. In particular, he says that

> as distinct from verbal art [i.e., literature], rhetoric, owing simply to its teleology, is less free in its handling of other speakers' utterances. Rhetoric requires a distinct cognizance of the boundaries of reported speech. It is marked by an acute awareness of property rights and by fastidiousness in matters of authenticity. (Voloshinov [Bakhtin] 122)

As the analysis that follows in the text should illustrate, however, the handling of reported speech in rhetorical texts is open to all the inflections of voices that Bakhtin discusses with regard to literature.

5. In Bakhtin's terms, this sort of ethos-building inoculation provides "hierarchical eminence" for the speakers, making their words much less susceptible to undermining. "The stronger the feeling of hierarchical eminence in another's utterance, the more sharply defined will its boundaries be, and the less accessible will it be to penetration by retorting and commenting tendencies from outside" (Voloshinov [Bakhtin] 123).

6. For the epistemological angle on this technique, see the enlightening discussion in Potter about the thoroughness with which "the attribution of interest is bound up with the construction and destruction of factual accounts" (55; see, especially 122–49; compare James 91–105, from which one of this essay's epigraphs comes).

7. Killingsworth and Palmer (15) discuss Carson's implied audience in similar terms, though they lament her willingness to spend all her suasive energy on "the presumably neutral and poorly informed general public" (i.e., the Citizens) at the complete expense of "the other side" (i.e., the Bad Guys). They may be right in the terms of their (much broader) perspective, from which they argue that such stances lead primarily to antagonism, though I am unconvinced that *Silent Spring* could have achieved its goals without the demonization it pursues.

8. See also p. 240, where Carson brings in Dr. Hueper to answer some hypothetical questions (which, presumably, she has asked him on behalf of her readers, in her role as mediator), and p. 274, where she herself again answers directly as an expert.

9. Also "systemic pesticides" (32), "penta" (36), "mutagens" (36), "biological magnifiers" (108), "moot" (159), "oxidation product" (170), "ginger paralysis" (197), "chemical carcinogen" (225), "male sterilization" (279), "juvenile hormones" (285), "gyplure" (296).

10. Also "draw the teeth" (32), "cock of the plains" (65), "the Reichenstein disease" (223), "wormy apples" (254).

11. In "Discourse in the Novel," Bakhtin discusses a phenomenon he terms "the 'word on display'" (322), which is a pretty good description of what scare quotes do to a word or phrase. Caryl Emerson and Michael Holquist suggest a word in quotation marks as their only example of this phenomenon in their glossary of Bakhtinian terminology (in Bakhtin, *Dialogic* 427). But I have seen nothing else in Bakhtin that comes this close to treating this currently rampant quotation-mark convention.

12. See Nash (79) for some discussion of this position with respect to Carson.

13. In addition to the quotation marks, she sometimes comments explicitly on her own language; for instance, "[T]he analogy is more poetic than precise" (200).

Works Cited

Aristotle. *On Rhetoric: A Theory of Civic Discourse.* Trans. George A. Kennedy. New York: Oxford UP, 1991.

Bakhtin, Mikhail. See also Voloshinov, V. N.

Bakhtin, Mikhail. "Discourse in the Novel." 1934–35. *The Dialogic Imagina-*

tion: Four Essays. Trans. Caryl Emerson and Michael Holquist. Ed. Michael Holquist. Austin: U of Texas P, 1981.

———. *Problems of Dostoevsky's Poetics.* 1929. Trans. R. W. Rotsel. Ann Arbor: Ardis, 1973.

Booth, Wayne C. *Modern Dogma and the Rhetoric of Assent.* Chicago: U of Chicago P, 1974.

Carson, Rachel. *Silent Spring.* 1962. Introduction by Al Gore. Boston: Houghton, 1994.

Fahnestock, Jeanne. "Accommodating Science." *Written Communication* 3 (1986): 275–96.

Gilbert, G. Nigel, and Michael Mulkay. *Opening Pandora's Box: A Sociological Analysis of Scientists Discourse.* Cambridge, UK: Cambridge UP, 1984.

Jacobinzer, Harold, and Harry W. Raybin. "Poisoning by Endrin." *New York State Journal of Medicine* 59 (15 May 1959): 2017–22.

James, William. *Pragmatism.* 1907. Ed. Bruce Kuklick. Indianapolis: Hackett, 1981.

Killingsworth, M. Jimmie, and Jacqueline S. Palmer. "The Discourse of 'Environmentalist Hysteria.'" *Quarterly Journal of Speech* 81 (1995): 1–19.

Myers, Greg. *Writing Biology.* Madison: U of Wisconsin P, 1991.

Nash, Roderick Frazier. *The Rights of Nature: A History of Environmental Ethics.* Madison: U of Wisconsin P, 1989.

Potter, Jonathan. *Representing Reality: Discourse, Rhetoric and Social Construction.* London: Sage, 1996.

Voloshinov, V. N. [Mikhail Bakhtin]. *Marxism and the Philosophy of Language: Basic Problems in Sociolinguistics.* 1929. Trans. Ladislav Matejka and I. R. Titunik. New York: Seminar, 1973.

Woolgar, Steve. *Science: The Very Idea.* London: Tavistock, 1988.

8

CHERYLL GLOTFELTY

Cold War, *Silent Spring:* The Trope of War in Modern Environmentalism

1962: Rachel Carson's *Silent Spring* is published. The Cuban missile crisis marks the height of U.S.– Soviet tensions, bringing the world to the brink of nuclear war.

๕๛ In 1962, after *Silent Spring* was serialized, one reader wrote to the *New Yorker,*

Miss Rachel Carson's reference to the selfishness of insecticide manufacturers probably reflects her Communist sympathies, like a lot of our writers these days.

We can live without birds and animals, but, as the current market slump shows, we cannot live without business.

An earlier version of this essay appeared in the 1995 *Proceedings of the Conference on Communication and Our Environment.* Edited by David B. Sachsman, Kandice Salomone, and Susan Senecah. Used by permission of the author.

> As for insects, isn't it just like a woman to be scared to death of a few little bugs! As long as we have the H-bomb everything will be O.K.[1]

As the chronology presented in the appendix and the above letter demonstrate, the genesis, the writing, and even the reception of Rachel Carson's most influential book exactly coincided with the Cold War years in America and were colored by them.[2] Indeed, the very subject of her book—DDT and other synthetic pesticides—was itself a product of war. DDT was first used on a large scale in the Naples typhus epidemic of 1943–44 and continued to be used during the rest of World War II to protect millions of soldiers and civilians against insect-borne diseases. Thanks to DDT, World War II is thought to be the first major war in which more people died from enemy action than from disease.[3] DDT, the miracle chemical of World War II, came to the United States for civilian use in 1945 in a wave of publicity and high hopes. Some called it the atomic bomb of insecticides, the harbinger of a new age in insect control. DDT and its fellow chemicals were heroes, fruits of a new age of science and technology that promised to make life more safe, comfortable, and convenient than ever before.

Causing a remarkable about-face in public opinion, the rhetoric of *Silent Spring* persuaded the public that these miracle pesticides were, in fact, deadly poisons, harmful to *all* living things, just as earlier Cold War rhetoric had convinced the American public that their World War II ally, the Soviet Union, had become their new worst enemy. After President Harry Truman delivered the speech that introduced what would become known as the Truman Doctrine, urging that the United States assist the democratic nations of Europe in their struggle against the powerful forces of totalitarianism, one commentator noted, "We went to sleep in one world and woke up . . . in another" (qtd. in Hinds and Windt 147).

Silent Spring had a similarly alarming effect upon public consciousness. It is featured in Robert Downs's *Books That Changed
America* as the book that launched the modern environmental movement. Scholars of American environmental rhetoric M. Jimmie Killingsworth and Jacqueline S. Palmer assert that "Carson's book established rhetorical conventions that would become standard fare
in the environmentalist debate" (65). *Silent Spring's* critique of the
widespread use of what were regarded as wonder chemicals was
nothing less than an indictment of modern life itself. Carson's editor and biographer Paul Brooks remarks, "*Silent Spring* has been
recognized throughout the world as one of those rare books that
change the course of history—not through incitement to war or
violent revolution, but by altering the direction of man's thinking"
(*House* 227).

I agree with Brooks that Carson altered the direction of our
thinking, but I argue that *Silent Spring* did instigate a new kind of
war by redirecting the language and concepts of the Cold War to
apply to "man's war against nature" (Carson 7). While pesticide
manufacturers were waging their all-out war on insects, Carson
started a new "war" against the manufacturers and pesticide spray-
ers. They became the new worst enemy. Industry as the enemy has
by now become an entrenched way of thinking in the environmental movement, and the lexicon of war continues to pervade environmentalist discourse. In this essay, I examine the way that Carson employs Cold War rhetoric in her effort to mobilize support for
a protracted war on despoilers of the environment.[4]

Pesticide advertisements themselves cleverly exploited Cold War
icons. Some pesticide commercials bore a striking resemblance to
anti-Communist propaganda, depicting an insect army marching
across a map of the United States. The situation looks terribly grim
until a spray can appears from the sky to kill the hostile invaders
dead in their tracks and keep the world safe for democracy, "sprout-

less potato[es]," and "mosquitoless patio[s]" (Carson 216).[5] Such advertisements equated pest control with national defense and made the spray can a symbol of patriotism.

Given the militant cast of pesticide advertisements, coupled with her own sympathies for the natural world, Carson might have chosen to adopt the strategy of arguing that insects and humans were *not* enemies and that war was the wrong metaphor to describe their relationship. To deconstruct the war metaphor as it was commonly applied to insects would have been a profoundly revolutionary move. Today's effort to reintroduce wolves on public lands, for example, has taken such an approach, reeducating the public to understand that wolves are not our enemies but are noble creatures whose presence enriches our lives. Carson, however, made only limited attempts to undermine the war metaphor by pointing out that there are good insects as well as bad insects and by questioning terms such as *weed* and *pest*. Her more dominant strategy is to agree with public sentiment that certain insects are indeed our enemies. Perhaps she sensed that it would have been impossible to persuade the public that gypsy moths, fire ants, and body lice are noble beings. She may have recognized that no amount of reeducation would convince a cotton grower that the boll weevil was not his enemy and that an attempt to do so would discredit her authority and alienate her audience.

Leaving the war metaphor intact, Carson's strategy is to speak as a smart general, insisting that we are making some critical errors in the way that we are fighting this war, errors that may prove to be suicidal or that could cause the insects to win. One such error occurs, she says, when we turn powerful weapons against ourselves. The midwestern states, for example,

have launched an attack [on the Japanese beetle] worthy of the most deadly enemy instead of only a moderately destructive insect, employing the most dangerous chemicals distributed in a manner that exposes large numbers of people, their domestic animals, and all wildlife to the poison intended for the beetle. (Carson 88)

Another critical mistake is made repeatedly when our effort to eradicate a species backfires and actually strengthens that species, which rapidly evolves to become resistant to the chemical. While we are thus encouraging the evolution of superpests against which further chemical attacks are futile, our indiscriminate chemical barrage is simultaneously killing the natural predators of the pest species. As Carson writes, "[W]e have turned our artillery against our friends. The terrible danger is that we have grossly underestimated their value in keeping at bay a dark tide of enemies that, without their help, can overrun us" (251). Carson thus concurs with an expert in soil science that "'a few false moves on the part of man . . . and the arthropods may well take over'" (qtd. in Carson 61). As these dire warnings imply, Carson does not idealize insects but depicts them as a force to be reckoned with; consequently, we had better learn to fight more intelligently and forge some alliances.

Carson not only exposes the folly of prevailing insect-control strategies, she offers numerous smarter battle plans, including the importation of a given pest's natural predators, introduction of insect-specific diseases, release of sterilized males, various kinds of lures and traps, improved sanitation, new strains of pest-resistant crops, and greater diversity in plantings. She concludes, "There is, then, a whole battery of armaments available to the forester who is willing to look for permanent solutions that preserve and strengthen the natural relations in the forest" (296). Carson promises no single solution, no atom bomb for insects, but rather recommends a variety of approaches, each depending on detailed, scientific knowledge of natural systems and each tailored to a specific situation. Just as America could not wipe out communism from the face of the earth but instead pursued a policy of containment and learned to establish diplomatic relations with a variety of Communist governments in many countries, so, too, Carson proposes that we seek not total extermination but rather "contain[ment] within reasonable bounds" (117) and "a reasonable accommodation between the insect hordes and ourselves" (296).

The foregoing discussion shows how Carson retains war-inflected language to propose a wiser defense plan against problem insects. War provides a readily comprehensible lens through which to view the relationship between humans and insects. The key is not to deny the war but to fight it more sensibly, with less risk to ourselves and our allies. While Carson thus agrees that certain insects are destructive, she argues that those very pests are less destructive to us than has been our chemical assault on them. Blanket spraying of potent synthetic chemicals poisons both humans and the natural systems upon which human life depends. Whoever is behind this rain of poisons is endangering our lives to a much greater degree than did the pest insects. By this clever move, Carson implies that the most deadly enemies are not the foreign insects but the insect controllers in our very midst. Such an accusation is tantamount to declaring a new war, a war between the pesticide industry and the people.

There are many striking parallels between the rhetorics of the Cold War, *Silent Spring,* and modern environmentalism. Cold War rhetoric constructed a bipolar world, one in which two superpowers dominated the globe, with smaller countries falling under one or the other sphere of influence. Differences within each camp were minimized, while differences between the two camps were exaggerated. The possibility of a third, alternative position was denied. With only two choices, the question becomes simply, Which side are you on? As is typical of wartime rhetoric, the "other" side is portrayed as malevolent and aggressive, while "our" side is innocent and defensive. Scholars of Cold War rhetoric Lynn Boyd Hinds and Theodore Otto Windt, Jr., have documented how "the Soviets

[were] transformed in this rhetoric of nihilation into subhuman monsters devoid of human feelings" (114). Hinds and Windt explain that melodrama is a staple of wartime rhetoric, as a sinister villain menaces a helpless maiden who is in desperate need of a stalwart hero (see also Kaldor; Medhurst et al.).

Silent Spring likewise creates a bipolar, melodramatic picture, with the pesticide industry and its henchmen in the Department of Agriculture on one side; Carson and a few heroic biologists and concerned citizens on the other; and with fainting nature and the unsuspecting American public costarring as damsels in distress. As Killingsworth and Palmer note,

> The agonistic exposé, of which *Silent Spring* is a fine example, . . . fosters controversy and divides perspectives, often attempting to arrange disparate interests into a clearly demarcated pair of opposed parties—environmentalist and developmentalist, for example—thus mobilizing citizens for a quick decision one way or the other. (76)

Carson depicts "the control men" (86) of the pesticide industry as inhuman, uncaring, and greedy. It is, in her words, "the authoritarian temporarily entrusted with power" (127) who orders the spraying and thereby "disregard[s] supposedly inviolate property rights of private citizens" (159) and "contaminate[s] the entire environment" (8). "On every hand there is evidence that those engaged in spraying operations exercise a ruthless power" (12), writes Carson. She observes that hers is "an era dominated by industry, in which the right to make a dollar at whatever cost is seldom challenged" (13) and in which "nothing must get in the way of the man with the spray gun" (85).

While the pesticide industry is clearly the villain, the public has been both a hapless victim of the "chemical death rain" (12) and an unwitting accomplice by passively allowing such abuse to happen.

Carson wrote *Silent Spring* in order to goad the passive to take action and to give the "many, many people who are eager to do something . . . the facts to *fight* with" (Freeman 391, emphasis added). The rhetoric of bad guys (commonly portrayed as a powerful corporate minority) versus good guys (portrayed as the increasingly vocal majority) is a familiar feature of modern environmentalism.[6]

During the Cold War, Winston Churchill's image of an Iron Curtain dividing Europe reinforced the sense of a bipolar world and, in addition, connoted that there were secret machinations taking place behind that Iron Curtain. As rhetoricians Hinds and Windt point out, this conception of mysterious dealings behind the curtain "pervaded interpretations of Soviet actions. Specific actions were never what they seemed, and one had to lift the masks from actors to interpret those actions for what they were" (107). In other words, Soviet statements might be lies, and Soviet actions concealed ulterior motives.

Carson adopts this Iron Curtain approach by questioning what deals are struck behind the closed doors of the Department of Agriculture, of university research labs funded by the chemical industry, and in the board rooms of pesticide manufacturers. She warns that "the control men in state and federal governments—and of course the chemical manufacturers— . . . steadfastly deny the facts reported by the biologists and declare they see little evidence of harm to wildlife" (86). Elsewhere, she refers to publications of the Department of Agriculture as "propaganda" (163). Sen. Joseph McCarthy's announcement in 1950 that he had a list of 205 Communists working in the State Department probably had nothing to do with Carson's insinuation that there are environmental traitors within government; nonetheless, both accusations kindled public mistrust of government, and similar indictments of government officials are by now a commonplace of environmental rhetoric.

During the Cold War, statements from Washington rarely described tensions with the Soviet Union in the context of historical development or national interests; instead, the conflict was cast in ideological terms as a war between communism and capitalism, between totalitarianism and freedom. The ideological chasm separating the two great powers was assumed to be unbridgeable. Ultimately, the Cold War became a moral struggle between good and evil, lightness and darkness. The mutual mistrust and persistent focus on incompatible ideologies all but precluded compromise, dooming negotiations about important, pragmatic issues like arms reduction.

Carson, too, cast what could have been a pragmatic discussion of appropriate pesticide use into the more philosophical and ideological context of the proper relationship between humanity and nature. Paul Brooks notes that Rachel Carson knew that she would be attacked: "It was not simply that she was opposing indiscriminate use of poisons but—more fundamentally—that she had made clear the basic irresponsibility of an industrialized, technological society toward the natural world" (Foreword xii). Carson insists that the pesticide issue has a moral dimension: "Incidents like the eastern Illinois spraying raise a question that is not only scientific but moral. The question is whether any civilization can wage relentless war on life without . . . losing the right to be called civilized" (99). As Carson represents this conflict, the unbridgeable ideological chasm is between domination and accommodation, arrogance and humility, stupidity and intelligence, greed and grace, right and wrong. Finding themselves publicly vilified, it is no wonder that chemical manufacturers attempted to block publication of the book and spent a quarter of a million dollars on a public-relations campaign. The ideological warfare of environmentalism continues today in much the same terms that Carson used, with a few new unbridgeable chasms appearing on the horizon, such as the ones separating patriarchy and ecofeminism, and anthropocentrism and biocentrism.

Finally, and perhaps most important, both the Cold War and

Silent Spring create a climate of crisis in order to justify their demand for drastic action. How was Truman to persuade Congress to spend millions of dollars to prop up corrupt governments like the one in Greece in 1947? He drew upon the public's memory of World War II to warn that, if Greece were to fall into the hands of the Communist revolutionaries, the rest of Europe could topple like dominoes. According to the rhetoric of the Cold War, as expressed by historian Louis Halle, the world stood at a crossroads in history: "'If the United States did not intervene now, all would be lost'" (qtd. in Hinds and Windt 134). Later, images of Armageddon were conjured up to pave the way for the quadrupling of the defense budget in 1950. In the blockade of Cuba in 1962, nothing less than national survival was said to be at stake. Hinds and Windt write, "In the medical world a crisis is a turning point for a sick patient, a crucial moment when a life or death decision must be made" (131). They observe that Cold War rhetoric created a series of crises until an atmosphere of crisis became the consensus reality.

Just as Truman drew analogies between the Soviet Union and the Nazis, Carson exploited the leading fear of her time—the threat of nuclear war—to draw parallels between pesticides and radioactive fallout.[7] The opening chapter of *Silent Spring* paints an imaginary picture of a pastoral American village suddenly afflicted with "a strange blight" that kills chickens, cattle, sheep, and even children at play. Carson then warns that a "grim specter has crept upon us almost unnoticed, and this imagined tragedy may easily become a stark reality we all shall know" (3). Elsewhere she observes that "anything . . . within range of the chemical fallout may know the sinister touch of the poison" (156). And because pesticides cause "a general and permanent lowering of environmental resistance," Carson predicts that "we may expect progressively more serious outbreaks of insects, both disease-carrying and crop-destroying species, in excess of anything we have ever known" (252). In fact, she says, "[I]t is happening, here and now" (252). She quotes a scientist, who confesses, "'We all live under the haunting fear that something may corrupt the environment to the point where man joins the dinosaurs as an obsolete form of life'" (188).

When, Carson asks, will the public demand action? Thanks to the wave of anxiety caused by *Silent Spring,* an answer would be come: In the summer of '62. While initially doubtful that she would be able to penetrate the barrier of public indifference to so uncharismatic an issue as pesticides, Carson succeeded in finding just the right rhetorical formula to galvanize American citizens. By working within the Cold War paradigm of her day, substituting the pesticide industry for the Communist bloc as the purveyor of impending doom, Carson was able to tap into a powerful source of public energy and emotion.

In part due to the precedent set by *Silent Spring,* the rhetoric of war, creating a climate of crisis, has been deployed by the environmental movement for more than three decades. One recent example from Dave Foreman's *Confessions of an Eco-Warrior* typifies the genre:

> We are living now in the most critical moment in the three-and-a-half-billion-year history of life on Earth. . . . Never before . . . has there been such a high rate of extinction as we are now witnessing. . . . Clearly, in such a time of *crisis,* the conservation *battle* is not one of merely *protecting* outdoor recreation opportunities, or a matter of aesthetics, or 'wise management and use' of natural resources. It is a *battle* for life itself. (2, emphases added)

The Cold War lasted for forty years. It will be interesting to see how long the consensus of environmental crisis can be maintained.

One goal of this essay has been to show how, having been catalyzed and colored by Rachel Carson's *Silent Spring,* the modern environmental movement has some of its roots in and continues to derive much of its rhetoric from war.[8] In some cases, the clash between environmentalists and anti-environmentalists has escalated beyond mere rhetoric, as author David Helvarg documents in *The War Against the Greens:*

> [I]t is hard for me to imagine that environmental conflict in the United States might ever begin to resemble some of the haunting scenes of violence and hatred I had come to know as a war correspondent in Northern Ireland and Central America. But today, . . . having seen the bomb and arson damage firsthand, and having met and talked to people who have been beaten, shot at and terrified, had their dogs mutilated, their cars run off the road, and their homes burned to the ground, I'm not so certain. (7)

In the heat of the environmental backlash, perhaps it is time to question whether the trope of war—with its battles, its victories and defeats, its ecowarriors and enemies, its moral crusades and mortal fear—is an appropriate tool for solving environmental problems and making intelligent decisions. Perhaps in Carson's day, war was a necessary and appropriate context in which to conceptualize environmental issues. But, thankfully, the Cold War is over. Should people who are committed to enlightened stewardship of the earth continue to invoke it?

Appendix
Chronology

1945	The United States drops atomic bombs on Hiroshima and Nagasaki, ending World War II.	Rachel Carson writes a letter to *Reader's Digest,* proposing to write an article on the harmful effects of wide-area applications of DDT.
1946	Joseph Stalin proclaims that war is inevitable as long as capitalism exists. In his "iron curtain" speech, Winston Churchill warns that the Soviets desire "the indefinite expansion of their power and doctrines" ("Mr. Churchill's Address"; see also LaFeber).	Rachel Carson is employed by the U.S. Fish and Wildlife Service.
1947	The Truman Doctrine pledges aid to democratic nations in their struggle against totalitarian regimes.	
1949	The Soviet Union explodes its first nuclear device.	
1950	The United States quadruples its defense spending. President Truman orders the development of the hydrogen bomb.	
1958		Carson receives a letter from her friend Olga Owens Huckins describing the bird deaths in her yard following the aerial spraying of her neighborhood with DDT. Carson determines to write a book about pesticides (see Brooks, *House*).

1959	Nikita Khrushchev forms an alliance with Fidel Castro.	Carson writes her editor, Paul Brooks, that she is building up "a really damning case against the use of . . . chemicals as they are now inflicted upon us" (Brooks, *House* 244).
1960	Summit talks in Paris between Khrushchev, Harold Macmillan, Dwight Eisenhower, and Charles de Gaulle fail.	Carson writes to her friend Marjorie Spock that "it is a great problem to know how to penetrate the barrier of public indifference and unwillingness to look at unpleasant facts that might have to be dealt with if one recognized their existence" (Brooks, *House* 258).
1961	President Kennedy takes responsibility for the Bay of Pigs invasion fiasco. Khrushchev orders construction of the Berlin Wall.	The title *Silent Spring* is chosen for Carson's nearly finished book. One working title had been *The War Against Nature;* another had been *Man Against the Earth* (see Freeman 286, 380).
1962	The Cuban missile crisis marks the height of U.S.-Soviet tensions, bringing the world to the brink of nuclear war.	*Silent Spring* is serialized in the *New Yorker* and then published in book form.

Notes

1. This letter, dated 29 June 1962, was sent to the *New Yorker* by H. Davidson of San Francisco. It was not published at the time, but it does appear in the "Better Late Than Never" section of the seventieth anniversary issue of the *New Yorker*. For an excellent report on *Silent Spring's* public reception, see Graham.

2. In *Rachel Carson: Witness for Nature,* Linda Lear documents how the Cold War formed the backdrop of Carson's entire career. Lear discusses military action in Korea, the specter of atomic warfare, *Sputnik I* and *II*, the Greater St. Louis Citizens' Committee for Nuclear Information findings of strontium 90 in baby teeth and cow's milk, the U-2 spy plane incident, the Cuban missile

crisis, and bomb shelter construction in America. It might also be noted that John Le Carré's novel *The Spy Who Came in from the Cold* appeared in 1964, shortly after publication of *Silent Spring*.

3. See Dunlap for a history of DDT; and for a good review of public perceptions and use of DDT, see the PBS television documentary "Rachel Carson's *Silent Spring*."

4. George Lakoff and Mark Johnson observe in *Metaphors We Live By* that war is one of the metaphors that our culture uses to conceive of experience and to structure discourse (4). Thus, in adopting the metaphor of war for rhetorical purposes, Carson is not inventing a new strategy but is following an already entrenched mode of expression. In this case, however, Carson's rhetoric of war is particularized and rendered all the more effective by the historical moment of the Cold War. As Lear notes, "In *Silent Spring*, Carson deliberately employed the rhetoric of the Cold War and the tone of moral crisis to persuade her readers of the urgency of her message" (428).

5. As Edmund P. Russell III demonstrates in his excellent essay, "'Speaking of Annihilation,'" the pesticide industry began using slogans like "the war against insects" to describe their activities shortly after World War I, just as the military had employed metaphors of insect control to refer to the human enemy. Although Russell's study ends in 1945, the links between war and insect technology, institutions, and propaganda persisted beyond World War II, when the war in question became the Cold War between the United States and the Soviet Union. See also Russell's Ph.D. dissertation, which examines the relationship between warfare and insecticides in the United States from 1879 to 1945.

6. See Randy Harris's essay, this volume, for an extended analysis of how Carson represents "Good Guys," "Bad Guys," and "Citizens."

7. For an interesting essay tracing Carson's effective allusions to radioactive fallout and extensive use of the language of radioactivity, see Ralph Lutts's 1985 essay "Chemical Fallout" (see also Lutts, this volume). Carson was not above putting the Nazis to work for her cause, either. For instance, she regularly draws attention to the German roots of the insecticides (as when she points out that dieldrin was named for a German chemist and when she identifies another insecticide as the product of German nerve gas research). My thanks to Randy Harris for pointing out Carson's use of Nazis.

8. See Dave Foreman and Bill Haywood's *Ecodefense: A Field Guide to Monkeywrenching* (complete with a camouflage cover) for an example of a handbook of tactical "field maneuvers" to use in the ongoing war to defend "Mother Earth."

Works Cited

"Better Late than Never." *New Yorker* 20 and 27 Feb. 1995.

Brooks, Paul. Foreword. *Silent Spring*. By Rachel Carson. 25th anniv. ed. Boston: Houghton, 1987.

——. *The House of Life: Rachel Carson at Work*. Boston: Houghton, 1972.

Carson, Rachel. *Silent Spring*. 1962. 25th anniv. ed. Foreword by Paul Brooks. Boston: Houghton, 1987.

Downs, Robert B. *Books That Changed America*. New York: Macmillan, 1970.

Dunlap, Thomas R. *DDT: Scientists, Citizens, and Public Policy*. Princeton: Princeton UP, 1981.

Foreman, Dave. *Confessions of an Eco-Warrior*. New York: Harmony, 1991.

Foreman, Dave, and Bill Haywood, eds. *Ecodefense: A Field Guide to Monkeywrenching*. 2nd ed. Tucson: Ludd, 1987.

Freeman, Martha, ed. *Always, Rachel: The Letters of Rachel Carson and Dorothy Freeman, 1952–1964*. Boston: Beacon, 1995.

Graham, Frank, Jr. *Since Silent Spring*. Boston: Houghton, 1970.

Helvarg, David. *The War Against the Greens: The "Wise-Use" Movement, the New Right, and Anti-Environmental Violence*. San Francisco: Sierra Club, 1994.

Hinds, Lynn Boyd, and Theodore Otto Windt, Jr. *The Cold War as Rhetoric: The Beginnings, 1945–1950*. New York: Praeger, 1991.

Kaldor, Mary. *The Imaginary War: Understanding the East-West Conflict*. Cambridge: Blackwell, 1990.

Killingsworth, M. Jimmie, and Jacqueline S. Palmer. *Ecospeak: Rhetoric and Environmental Politics in America*. Carbondale: Southern Illinois UP, 1992.

LaFeber, Walter. *America, Russia, and the Cold War, 1945–1980*. 4th ed. New York: Wiley, 1980.

Lakoff, George, and Mark Johnson. *Metaphors We Live By*. Chicago: U of Chicago P, 1980.

Lear, Linda. *Rachel Carson: Witness for Nature*. New York: Holt, 1997.

Lutts, Ralph H. "Chemical Fallout: Rachel Carson's *Silent Spring*, Radioactive Fallout, and the Environmental Movement." *Environmental Review* 9.3 (Fall 1985): 211–25.

Medhurst, Martin J., Robert L. Ivie, Philip Wander, and Robert L. Scott. *Cold War Rhetoric: Strategy, Metaphor, and Ideology*. New York: Greenwood, 1990.

"Mr. Churchill's Address Calling for United Effort for World Peace." *New York*

Times 6 Mar. 1946, late ed.: 4. [Text of address delivered at Westminster College, Fulton, MO.]

"Rachel Carson's *Silent Spring.*" *The American Experience*. Writ. and prod. Neil Goodwin. PBS. Peace River Films, Cambridge. 8 Feb. 1993.

Russell, Edmund P., III. "'Speaking of Annihilation': Mobilizing for War against Human and Insect Enemies, 1914–1945." *Journal of American History* 82.4 (1996): 1505–29.

———. "War on Insects: Warfare, Insecticides, and Environmental Change in the United States, 1879–1945." Diss. U of Michigan, 1993.

9

M. JIMMIE KILLINGSWORTH AND
JACQUELINE S. PALMER

Silent Spring and Science Fiction: An Essay in the History and Rhetoric of Narrative

The aim of science is to discover and illuminate truth. And that, I take it, is the aim of literature, whether biography or history or fiction; it seems to me, then, that [there] can be no separate literature of science.

—Rachel Carson, National Book Award Acceptance Speech

After all, the boundaries between fiction and nonfiction, between literature and nonliterature and so forth are not laid up in heaven.

—M. M. Bakhtin, *Dialogic Imagination*

Hubris clobbered by nemesis.

—Brian W. Aldiss, *Trillion Year Spree*

❧ By exploring a connection between Carson and science fiction in the history of environmental literature, we are risking alignment with the earliest critics of *Silent Spring*. One scientist, defending the pesticide industry in a review of Carson's book, wrote dismissively, "I regard it as science fiction, to be read in the same way that the

174

TV program 'Twilight Zone' is to be watched" (qtd. in Graham 39; see also Bosso 117). For this critic and for others, "science fiction" sums up the kind of writing that real science rejects—the speculative, emotional, sensational products of the entertainment industry, utterly devoid of the rigor, objectivity, and clarity of true scientific writing.

While we would never countenance a dismissal of Carson along such lines nor seriously suggest that Carson was herself a science fiction writer, we contend that the literary parallels between *Silent Spring* and science fiction, both in form and content, are worthy of our attention.[1] One thing that all of Rachel Carson's writings share with science fiction is that they testify to the ultimate superficiality of the "two cultures" theme, so winningly articulated by C. P. Snow. The theme has held Western intellectuals in a spell of fascination for nearly half a century and has driven into neglect many instances of "cultural interchange" between science and literature. One result is that the literary presentation of science for a nonspecialized audience—science *nonfiction* literature, to which Rachel Carson devoted nearly her entire career—has been all but totally neglected by historians of literary nonfiction. As for science fiction, the dismissive attitude of Carson's reviewer is typical not only of the scientific establishment but also of most professors of literature; it is reduced to a topic of "popular culture." We suspect, however, that as a literature of recruitment—writing that fires the imagination and inspires interest in scientific topics—science fiction shares an honored place with popular science nonfiction, a position that no science textbook, specialized article, or university lecture could fill.

Our aim is to take science fiction seriously, not as a substitute for science, but as a response to science that contributes to the development of *myths*. And we treat myths not as the cultural equivalent of lies, mistakes, or superstitions that scientific enlightenment is committed to destroy but as collective narratives reaching beyond the boundaries of any specialized body of knowledge and touching the heart of a society's emotional, spiritual, and intellectual consciousness. Even for science, myths can have a formative effect, working at times below the level of objective consciousness, as unconscious or semiconscious influences.

We follow the lead of Christine Oravec's study in this volume, which suggests that, as a result of Carson's meticulous revisions, the opening chapter of *Silent Spring* has a mythic force. Since myths have a narrative structure, to invoke a myth within an argument, or to attempt to create one, a writer must tell a story, and Rachel Carson was a powerful storyteller. We contend that not only the opening chapter, "A Fable for Tomorrow," but *Silent Spring* as a whole has a narrative thrust. Moreover, within every chapter, numerous stories unfold to fill out and extend the general outline of the myth developed in "A Fable." We begin by examining some of the rhetorical advantages, as well as some of the problems, that Carson's opening narrative creates in light of similar problems encountered by the energetic mythmakers of modern science fiction. We show how she taps into a vein of mythology that has haunted literature since the Enlightenment, the story of the end of the world brought about by human hubris, the counternarrative to the master story of human progress and perfectibility. We then explore the development of competing narratives within the whole of *Silent Spring* and show how the history of science fiction following 1962—possibly under Carson's influence—relives again and again the narrative conflicts inaugurated in *Silent Spring*. The difficulty of balancing apocalyptic warnings and millennial hope—and the mythic representations of both—continues to torment ecoactivist writing, as we demonstrate in our final section.

Our overall aim in this essay is to refine a key idea of our previous writings: Carson stands at the center of a literary tradition that includes most of the important nonfictional writing about the environment that has appeared since her time and that also includes much of the science fiction.[2] Our main methodological claim is that, by studying myths and narratives of all kinds across the lines imposed by specialization and genre, we enrich our understanding of narrative possibilities and in the best case create new options for thought and action. The differences between fiction and nonfiction are not "laid up in heaven," as Bakhtin wittily remarks (33), but are socially constructed, often in an attempt to deny one mythology, or one cultural perspective, in favor of another. Our purpose as discourse critics is to document similarities (or differences) that are denied and try to get at the reasons for the denial in hopes of producing insights about the stakes of discourse formation on topics as important as the human relation to the natural environment.

"A Fable for Tomorrow" as Science Fiction: The Apocalyptic Overture

Though Carson never intended it as such, the opening chapter of *Silent Spring*, "A Fable for Tomorrow," can be read as a brief experiment in science fiction. It is a narrative of the future that presents technological development in terms of the magical and the fantastic, confronting modern science with its old nemesis and competitor—witchcraft—even suggesting that science can devolve into witchery, can produce killing "spells," and that a people unaware of the side effects of the very powers that bring them prosperity, spellbound by their own success, can end by permitting the destruction of what they most love, first the grass and trees, next the birds of the air and beasts of the field, and finally themselves and their children, their hope for a future. The claim at the beginning of the tale about a fictional "town in the heart of America" on which "some evil spell had settled" sets up the fine, ironic twist toward the end of the story: "No witchcraft, no enemy action had silenced the rebirth of new life in this stricken world. The people had done it themselves" (3).

It is a gripping plot and one that is well known to writers and readers of modern science fiction. It fits surprisingly well the influential, if controversial, definition of science fiction provided by Brian W. Aldiss: *"Science fiction is the search for a definition of mankind and his status in the universe which will stand in our advanced but confused state of knowledge (science), and is characteristically cast in the Gothic or post-Gothic mode"* (25, his emphasis). It fits even better the whimsical short version of Aldiss's definition: *"Hubris clobbered by nemesis"* (26, his emphasis).

In the decision to follow the apocalyptic story line and in the use of the language and imagery of witchery and magic, which in science fiction has always shadowed the language of enlightened progress, Carson creates a narrative that resonates with the Gothic mode of modern science fiction. As with Mary Shelley's portrayal of Victor Frankenstein, in what is arguably the first science fiction novel (Aldiss 18), scientific humanity is overwhelmed by its own power to control nature, and the Fall of humankind remains a possibility even as science seeks human perfectibility. In Carson's "Fable for Tomorrow," progress is inverted; the human attempt to control nature, to improve upon nature, leads finally to the death of nature: hubris clobbered by nemesis indeed.

In trying to extract a warning with the apocalyptic narrative, however, Carson was courting a number of potential difficulties. In its modern form, the apocalyptic tale ends not with the coming of a new kingdom at the end of the natural world, as in the Christian apocalypse, but with the simple end of the human race. True millennialists see the end of the world—or, more precisely, the end of an era represented in end-of-the-world language—as a cyclic phenomenon to be accepted as inevitable, a sign of the new age, something hopeful, to be anticipated with joy. By contrast, the modern apocalypse represents the utterly complete destruction of the human race, if not the whole earth (see Buell 301). Even in modern life, however, contemplating end-times can provide a somewhat obscure form of pleasure for readers, which probably suggests why the theme has frequently proved attractive to fiction writers. Science fiction authors were producing end-of-the-world narratives as early as H. G. Wells, who in *The Time Machine* portrayed an earth at the end of time reduced to a state of inorganic depletion. In the age of the atom bomb, the causal responsibility for the end shifted from the hand of God or Nature to that of human beings. In all of its versions, however, the apocalyptic narrative has invited readers to reflect upon the place of humanity in the long history of life. Since modern natural history tends to minimize the status of the human race in the overall scheme of things, stories of the end of the world—no matter who or what initiates the course of action that brings about the end—provide a perverse comfort to individual humans discouraged by the prospect of their condition as traditionally portrayed, a hard life ending in death. The little life of the person now finds an allegorical representation in the relatively little life of the total human race. And modernism, the death of the old to make room for the new, likewise finds an analog in the myth of human extinction.

In addition to having to overcome the belief that scientific humankind will live forever, marching evermore certainly toward perfection, then, Carson may also have inadvertently invoked something like a collective death wish, the satisfying apocalypse, which would negate the power of the warning. And even if the warning were to get through this cognitive barrier, it still must pass another. The portrait of the future must not be taken literally as an inevitable fate but rather as a possible future that can be avoided.

The literalist response is one that science fiction writers and commentators struggle with continually. It comes from a tendency of readers to regard science fiction narratives as a predictive form of prophesy, to view them as *extrapolative*. In the introduction to her novel *The Left Hand of Darkness*, Ursula K. Le Guin describes precisely the technique Carson uses in "A Fable for Tomorrow" and the problem she faces as an author of narratives about the future. "Science fiction," says Le Guin, "is often described, and even defined, as extrapolative. The science fiction writer is supposed to take a trend or phenomenon of the here-and-now, purify and intensify it for dramatic effect, and extend it into the future."[3] Science fiction is said to resemble scientific practice in this way:

> A prediction is made. Method and results much resemble those of a scientist who feeds large doses of a purified and concentrated food additive to mice, in order to predict what may happen to people who eat it in small quantities for a long time. The outcome seems almost inevitably to be cancer.

This outcome is identical, ironically, to one of the outcomes contemplated in *Silent Spring*. Le Guin goes on to connect the method of extrapolation specifically with ecological prediction, not with Rachel Carson as it turns out—Carson in fact stopped short of making explicit the dire predictions implicit in her "Fable for Tomorrow"—but rather with the Club of Rome, an ecological think tank that did not balk at predicting Armageddon: "Strictly extrapolative works of science fiction arrive about where the Club of Rome arrives," says Le Guin, "somewhere between the gradual extinction of human liberty and the total extinction of terrestrial life."[4] When questioned closely, Le Guin observes, people who avoid science fiction—and the same could be said of people who avoid the apocalyptic variety of environmentalist narrative—avoid it not because it is "escapist" but because "it's so depressing."

Le Guin does not accept the extrapolative definition of science fiction exactly for this reason: just as it limits human freedom, the possibility that individual human beings can act ethically for the general good of the human and nonhuman world, it also limits in a parallel way the possible outcomes of the story—"Almost anything carried to its logical extreme becomes depressing, if not carcinogenic." Extrapolation thereby places undue limits on the narrative imagination. It is only one element in the futurist narratives of science fiction, Le Guin insists, one that is "far too rationalist and simplistic to satisfy the imaginative mind, whether the writer's or the reader's." Realizing the significant link between narrative technique and the ability to think through human problems and actions, Le Guin makes it her mission to break free of the limits that extrapolation imposes. She does so by thinking of science fiction not as predictive but as descriptive, not as future oriented but as reflecting the present, the now, thematically and psychologically, if not literally. Science fiction does not merely warn us of the ultimate dangers of cancer; it shows us our carcinogenic, as well as our more healthful, conditions in all of their dimensions and invites us to engage in *thought-experiments,* a form of meditation, again typical of science, but this time theoretical science instead of laboratory science: "The purpose of a thought-experiment, as the term was used by [Erwin] Schrödinger and other physicists," Le Guin argues, "is not to predict the future—indeed Schrödinger's most famous thought-experiment goes to show that the 'future,' on the quantum level, cannot be predicted—but to describe reality, the present world." The best of science fiction can be interpreted and written as a thought-experiment: "In a story so conceived, the moral complexity proper to the modern novel need not be sacrificed, nor is there any built-in dead end; thought and intuition can move freely within bounds set only by the terms of the experiment, which may be very large indeed."

Reading *Silent Spring* and the literature of environmental activism inspired by Rachel Carson alongside the literary history of science fiction, we have come to the conclusion that the stories activists tell about how people relate to the environment—as well as

science fiction stories explicitly treating environmental themes—
are subject to the same narrative limitations Le Guin describes in
her discussion of extrapolation. Too often obsessed with prediction,
overly rationalistic, and simplistic in plot design, they represent a
major problem for the environmental imagination: how to break
free of the master plots that confine our understanding of humanity's relation to the world while at the same time appealing to an
understanding of the world common enough to inspire the assent
of the reading public. In reading a story, a person should be able to
say both "This world is recognizable" and "I hadn't thought of
that." Only through such narratives can both communal recogni-
tion and commitment to change coexist. *Silent Spring* is unique and
beautiful because it is just such a narrative. But readers of *Silent
Spring*—especially the critical and dismissive readers but, as we
demonstrate, even some enthusiastic readers and writers influenced by Carson's work—have too often failed to respond sensi-
tively to the totality of the vision it presents. Ironically, this failure
may be the direct result of the power of Carson's opening narrative
to take hold of the reader's imagination.

Part of the power comes from Carson's conception of her introductory story as a fable, which suggests another connection with
science fiction. In Robert Scholes's provocative formulation, the
popular abbreviation *sf* for science fiction is translated into "structural fabulation," in which "the tradition of speculative fiction"—
Robert Heinlein's version of sf—"is modified by an awareness of the
nature of the universe as a system of systems, a structure of structures, and the insights of the past century of science are accepted
as fictional points of departure" in explorations of "human situations made perceptible by the implications of recent science"
(Scholes 41–42). Carson's version of ecology certainly provided her
with a "system of systems." The idea of using the fable as a structural device may have been suggested to Carson by her love of animal fables, a genre of narrative that, like science fiction, uses dis-

placement, defamiliarization, and the fantastic to create a kind of ironic distance that would not be possible in a realistic story. We can laugh at the manic Mr. Toad in *The Wind in the Willows,* but the laughter gives way to moral discomfiture as our realization grows that his acquisitiveness and frenetic grasping are all too typical of people in our age. *The Wind in the Willows* was among Carson's favorite books. In her published correspondence with Dorothy Freeman, she mentions it with great admiration five times. By contrast, she mentions Thoreau only twice. She seems to have thought of Thoreau in the same way many people think of the Bible or of Winston Churchill, primarily as a bank from which to draw quotations and maxims, whereas *The Wind in the Willows* and other "children's books," notably *Alice in Wonderland,* she engaged with real passion and interest and indeed questioned whether they were really for children after all (see Freeman 83, 85, 177, 266, 288, 322, 336). As Lawrence Buell has suggested, Carson's writing presents the reader with a world that because of the "witchery" of human action "threatens to turn monstrous" (291). Carson herself was perfectly aware of the connection of her rhetoric to the fabulous and the fantastic: "The world of systemic insecticides is a weird world," she said in *Silent Spring,* "surpassing the imaginings of the brothers Grimm" (32).

What Carson did not realize was that, in drawing upon the apocalyptic tradition to create a cautionary fable, she had to overcome many distracting resonances before her full message could be effectively delivered. She wanted her readers to take warning and take action, not just revert to contemplating (with either pleasure or depression) the inevitable end of life. "A Fable for Tomorrow" created for many of the earliest readers of *Silent Spring* an impression best described as *stunning.* Not only did Carson attract more attention and create more alarm than she could have imagined; she also selected a rhetorical structure that condemned many of her best and most positive points to relative neglect.

The "Fable" works as a thematic overture for the book as a whole. Its narrative threads reappear later as more specific, factual stories about the death of birds, the browning of grassy roadsides, and human illnesses and even deaths resulting from accidental poisonings in pesticide handling. However, the threads of narrative not anticipated by the "Fable" get lost in the apocalyptic jumble of the rest. Taken as a whole, *Silent Spring* favors a millennialist version of progress rather than a strictly modern apocalypse, which views the end of the world as absolute. This remake of Christian millennialism has in fact been the dominant allegory of industrial culture, the idea being that the Enlightenment was the dawn of the new millennium replacing the Dark Ages that followed the collapse of classical learning and the Roman Empire. But Carson displaces the myth in history. In her version, a "Neanderthal science" with powerful but crude instruments has driven industrialist change and has led civilization to the brink of a new Armageddon. And yet, in the new ecological sciences with their subtle grasp of interrelated systems, there is reason to hope. The millennialist language suggests the outlines of scientific history as portrayed in Thomas Kuhn's *Structure of Scientific Revolutions,* which was published in the same year as *Silent Spring.* Carson is furthering a Kuhnian revolution in science, a paradigm shift, an end to one narrative thread and a beginning for another, progress developing in waves rather than in one long, absolutely straight line. The stories of how ecologists first documented the troubles with industrial pesticides tend to support the apocalyptic vision early in the book—the end of one story of progress—and only in the second half does the suggestion appear that, having rooted out the problem, ecological wisdom may also lead to the solution, the beginning of a new era of human progress. Heroes abound in *Silent Spring*—not only wildlife biologists like Maurice Baker and Clarence Cottam, or cancer researchers like W. C. Hueper and Morton S. Biskind, but also ecologically informed entomologists like A. L. Melander, who hypothesized the notion of insect resistance to poisons decades before the development of DDT; or Edward Knipling, a pioneer in biological control of insects, who experimented with the release of sterilized males into a population of an offending species. The trouble is that the stories of successful biological control of insects and of effective guidance in the adaptive management of nature, Carson's first steps along the path of what would two decades later become known as the paradigm of sustainability, leave their mark too faintly in the overall narrative structure of *Silent Spring,* appearing too late in the book and without the added power of having been forecasted by her opening "Fable."

The Novelistic Momentum of *Silent Spring:* The Millennialist Turn

In *The Environmental Imagination,* Lawrence Buell writes that *Silent Spring* lacks the obvious elements of a novel—"no hero, no cast of characters, no narrative plot"—but the book's "disenchantment with the pesticide industry's witchery has a novelistic momentum to it, building from considerations of earth, water, and plants (Chapters 4–6) to wildlife (7–11), culminating in a chapter on cancer" before finally holding out the hope of a heroic recovery of environmental integrity under the influence of holistic ecology (293–94). We want to go a bit further and argue that much of the power of *Silent Spring* comes from the narrative momentum that Buell identifies and that a narrative rhetoric is at the heart of the book. Contrary to Buell's remark about the omissions of narrative elements, in fact, beyond the opening chapter, "A Fable for Tomorrow," a fairly definite plot and cast of characters, both heroes and villains, emerge. Buell may have overlooked these narrative features because, like many readers of *Silent Spring,* he focuses too closely upon the opening chapter in interpreting the book as a whole.

The master narrative that Carson takes as her theme is no less than the story of human progress.[5] Taken as a whole, *Silent Spring* offers both a critique and a revision of the progressive narrative. Carson's book appears to explore with great intensity all the discourse paths of the master narrative of progress in a moment of crisis. *Silent Spring* appeared at a time when the discourses of science nonfiction and science fiction were becoming increasingly critical of industrial society. It was the age of the Cold War; the world was still reeling from the production of the atomic bomb by an alliance of the Western scientific establishment with the United States military. It is no accident that *Silent Spring* employs the imagery of nuclear disaster explicitly in its critique of the "chemical death bombs" and "biocides" of the industries Carson condemns. As Linda Lear has suggested, "The technology that produced the atomic bomb gave humans the illusion of power. Now they had the ability to unleash forces which eventually would outrun their control"; and so it was that "For Carson, like many scientists of her generation, the atomic bomb changed forever the way she perceived the living world" (28). In the arena of science fiction, the same issues of power and control in the nuclear age inspired writers of Carson's generation, such as Frank Herbert, John Brunner, and Ursula Le Guin, to pull away from Hugo Gernsback's model of science fiction as a comic-book-hero approach to enlightening the public about scientific ideas and the future of technology (see Bainbridge; James). The new authors were transforming science fiction into a powerful literature of social criticism, drawing on earlier, primarily European models, such as the work of Mary Shelley, H. G. Wells, Aldous Huxley, and George Orwell.

The redirected aims of science fiction crisscrossed those of science nonfiction in the 1960s. In the literature of ecology, Carson was joined by writers like Paul Ehrlich and Barry Commoner in turning scientific skepticism back upon scientific practice itself. Though they built upon the tradition of earlier activists like Thoreau, Muir, and Leopold, they also broke the hold of the old contemplative nature essay as the primary medium for reflections about humanity's relationship to the natural world, using instead the modern genre of popular science writing for the first time as a literature of science-based activism. Before writing *Silent Spring,* Carson was a popularizer of the old school, an "educator" of a grateful public curious about science but left in the dark by the professionalization of scientific research and literature (see Lear). When it came to dealing with the threat of pesticides, however, an angrier author awakened. Carson turned her scientific training, as well as her reputation as a public educator, to the business of large-scale social protest and political influence. Her work, as she saw it, was to raise the awareness—this time the political and the scientific awareness—of the general citizen by putting together the fragments of knowledge generated by overly specialized professional science. As she wrote in *Silent Spring,* "There is still very limited awareness of the nature of the threat. This is an era of specialists, each of whom sees his own problem and is unaware of or intolerant of the larger frame into which it fits" (13). She had faith that, when people knew the whole story and could interpret its meaning for the long run of life on earth, these educated readers would respond with sympathy and outrage.

Even Buell, one of Carson's most sympathetic and careful interpreters, reads *Silent Spring* as an apocalyptic vision with a weakly positive millennialist conclusion. In Buell's view, "Carson takes us to the edge of catastrophe and then offers 'The Other Road'—without, however, offering much hope that it will be taken" (293–94). Most of the mainstream American writers named by Buell and another student of modern apocalyptic literature, Douglas Robinson—authors like John Barth, Kurt Vonnegut, Jr., Thomas Pynchon, and Margaret Atwood—make it "hard for apocalypticism to keep a straight face" (Buell 300). But Carson takes the possibility of the end of the world not merely as a trope for the end of an age but as a literal possibility. In this respect, Carson's work is more like that of the Native American writer Leslie Marmon Silko, according to Buell:

> The high seriousness of texts like Silko's [novel *Ceremony*] and Carson's [*Silent Spring*] brings out more strikingly the pastoral logic that undergirds environmental apocalypse, which rests on the appeal to the moral superiority of an antecedent state of existence when humankind was not at war with nature. (300–301)

While this assessment is certainly true of "A Fable for Tomorrow," which gives us an image of the happy village corrupted by modern ways, it misses both the pervasive modernism and the possibility of an alternative narrative in the later chapters of *Silent Spring*. There is no doubt that Carson is serious about the human ability to destroy nature and all of humankind along with it, but only in "A Fable" does she seem to look backward toward some ancient and ideal condition of life. By reading "A Fable for Tomorrow" as an overture that captures the very heart of Carson's argument, Buell asserts that "Carson grounds her appeal in a vision of the mythical American small community" (301). Rather than a grounding for this appeal, we would argue, Carson's memories of a lost past represent more a point of departure. The pastoral idyll tends to pass from view as Carson begins to consider alternatives and solutions and to present us with the heroic ecological vision in which her strongest claims are most firmly grounded. As Lear has suggested "'ecology' became part of everyday vocabulary" with the publication of *Silent Spring*, the "most enduring legacy" of which was a "new ethic of interconnectedness" (Lear 28, 42). With a web of interrelated stories, *Silent Spring* shows that the green fuse running through all life, including the genetic heritage of humankind itself, transmits all too readily the explosive poisons of systemic pesticides. The evidence of the poisoned elm–earthworm–robin cycle is only the first indication of a more sweeping and sinister problem. And the ecology of connectedness is no throwback to a sentimentally conceived ideal past, nor even a romantic ideal such as we find in the nineteenth-century poets or in Thoreau; it is a thoroughly modern conception, the biological equivalent of globalism in the political realm, the idea that no action is innocent of repercussions throughout the larger system, whether that system is an ecosystem or a political system. In embracing such a view, Rachel Carson proves herself a scientific realist and cuts herself off from the comfort of a pastoral nostalgia, though for readers struck most heavily by her "Fable for Tomorrow"—readers ranging from the industry parodists who produced counterfables like *The Desolate Year* (Monsanto Chemical Co.) and "Silent Autumn" (Dow Chemical) to those who continued to view her patronizingly as a timid bird-watcher who stumbled haltingly into the public storm of pesticide politics—Rachel Carson could be classed (and dismissed) among those early environmentalists whose main aim was to preserve the world of their lost childhood or their favorite hiking and camping spots.

"A Fable for Tomorrow" succeeded mightily in drawing the attention of a broad audience, but as a prologue or an overture, it failed to convey adequately the multiple thematic directions of the book. It is in fact not a prologue, but simply chapter 1, a point of departure. By chapter 7, we are already reading of solutions and alternatives to overapplication of pesticides. These moments of hope appear throughout the rest of the book and are brought together in the hopeful and passionate conclusion in chapter 17, "The Other Road."

Our appendix provides a chapter-by-chapter summary of the overall narrative thrust of *Silent Spring* along with samples of the short narratives—the cases and anecdotes and characters—that appear in each chapter. The characters include not only human heroes, villains, and victims but also important animal "characters" with symbolic and sympathetic force and personified figures like the pesticidal chemicals that appear as demonic characters in the story of the earth bewitched by human hubris.[6]

From this summary, the shape of the overall narrative becomes clear: while we were sleeping, as it were, the demon pesticides have spread among us, threatening our natural resources, our wildlife, our foodstuffs, and finally ourselves. Their presence was first invoked by industrial scientists, government agencies, and agribusiness, misled by their foolish philosophy that centered on the control of nature. The spread and maintenance of the demonic chemicals were abetted by the same agents with their self-deceiving propaganda but were resisted by holistic ecologists and by citizens gradually awakening to the destruction of their beloved world. The ecologists have offered, in addition to warnings, alternative means of solving the very real agricultural problems that originally prompted the use of chemical pesticides. We now await the outcome of the conflict.

Having taken on the apocalyptic narrative, as Buell suggests, Carson seems to have had difficulty summoning the energy to tell the story of "The Other Road" with its heroes sounding their warnings and posing alternative solutions, heroes who are mentioned fairly early (even in the acknowledgments, which actually precede chapter 1) but not fully displayed until chapter 17. If we read the book as a novel, which saves some of the most important information until the end, to keep the reader reading with suspense, or like an old-fashioned scientific paper, which reserves the most important conclusions until the last section, then chapter 17 appears as an emphatic rhetorical position. Very likely, however, many readers never make it that far. They have decided by then that the author is a prophet of doom and will have little to offer that is not depressing.

Thus, the conflicting narratives of apocalyptic doom and millennial hope strive for dominance in *Silent Spring*. In this, as in so many other ways, it is a culture-bearing book, gathering the threads of mythology that precede it, reweaving them, and casting into the future a narrative fabric that will become the fascination of a new generation of writers. A look at a few examples of the environmentally oriented science fiction that preceded and followed the publication of *Silent Spring* makes this trend clear.

Apocalyptic and Millennial Narratives: Parallels in Science Fiction

The difficulty Rachel Carson faced in envisioning, promoting, and sustaining a heroic millennialism against the allure of apocalyptic environmentalism reappears in the literary history of science fiction. Stories of alternative scientific paradigms and visions of aliens or forgotten tribes leading civilized humans to a new enlightenment cropped up early in the 1950s—in the *Foundation* novels of Isaac Asimov, for example, who later collaborated with Frederik Pohl to publish the nonfiction book *Our Angry Earth*; in Arthur C. Clarke's *Childhood's End*, which allegorizes human progress as the death of one way of life and the alien-assisted emergence of a new generation of children as leaders in the next stage of human evolution; and in Damon Knight's 1952 novella *Natural State*, in which a rural band of ostensibly ignorant hillbillies defend themselves and their advanced biotechnology against first the sales pitch and then the military action of an urban-based industrial society, finally defeating the urbanites and replacing their rusted and run-down mechanized world with a greener, biologically based technological culture.[7]

By the 1960s, in the wake of *Silent Spring*, the narrative of millennial ecology flourished in science fiction. The hero of Frank Herbert's *Dune* (1965) realizes his fate to become a great political leader only by joining with tribes of ecologically sophisticated natives of a desert planet who effectively resist mechanized extractive technologies in a galaxy full of industrialized civilizations. In similar fashion, Ursula Le Guin's 1969 novel *The Left Hand of Darkness* tells the plaintive but hopeful story of a desolate planet and the people whose culture is shaped by its conditions. Her protagonist, an offworld diplomat, successfully adapts to the ways of that world and completes his mission only by cultivating a mystical attitude of *presence* that complements his earthly hunger for progress. This tradition in science fiction culminates in Ernest Callenbach's 1975 *Ecotopia*, the tale of how a committed industrial progressivist adjusts to an alternative culture that becomes a nation of green activists after a bloodless revolution.

Unfortunately, the story of successful adaptation and alternative technologies has not continued to prevail as a narrative line in ecological literature, neither in science fiction nor in the discourse of environmental activism, which resorts again and again to sounding the apocalyptic alarm despite the rhetorical risks of this strategy. After *Ecotopia*, Ernest Callenbach wrote a weak sequel, then disappeared from the public scene. And after their early novels of heroic adaptation, which integrated environmentalist concern with more general cultural, political, and philosophical concerns, both Herbert and Le Guin isolated the theme of environmental impact and found themselves trapped in the facile dichotomy of industrial development versus environmental protection. In *The Green Brain* (1966)—a novel that clearly shows that the author was familiar with *Silent Spring* and the controversies surrounding it—Herbert imagines the insect world rising up against the global application of industrial pesticides, developing not only effective resistance in the physical sense but also a collective consciousness capable of reason, communication, and political resistance. As the worker insects make war upon developers in the rain forests of Brazil—who do their work against the will of a weak minority of environmentalists, disparaged as "Carsonites" and consigned to the background of the book—the green brain itself voices an appeal to readers to bring a halt to the war of human beings against nonhuman nature, a war that is not only destructive but also ultimately ineffective.

Along the same lines, though with the greater sensitivity and style that marks all of her work, Le Guin presents in *The Word for World Is Forest* (1972) the conflict of an extractive, industrially oriented culture of earthlings who undertake to colonize a new planet for the sake of logging so that wood, now a precious commodity on the earth, can be returned via space freight at a huge profit. The native humanoids are at first passive and even amenable to slavery but grow restless as their world and culture are increasingly destroyed, take up the foreign practice of murder and war, and finally destroy the operation of the invading profiteers and secure a truce that protects their environment and way of life. In her later novel, *Always Coming Home* (1985), Le Guin shows us a postindustrial tribe of earth people more like Native Americans than Americans of the industrial age in their culture and earth-binding practices. Centered in the pastoral tradition that Buell identifies, the millennial theme developed in these works takes on a cyclic shape: the industrial culture meets and overruns its limits and thereby returns the human race to a humbler relation with the nonhuman world. In Herbert's *Dune* and in the later novels of Le Guin, this means the triumph of tribal cultures, collective consciousness, and environmental integration over industry, Western capitalism, and the cultural pattern of individualism.[8]

As Herbert and Le Guin turned to a more darkly apocalyptic vision of the possible outcomes of modern industrial society, they joined the likes of environmental activist Paul Ehrlich, who not only dropped his *Population Bomb* upon the public in 1968 but in the next year wrote a science fiction version of his Malthusian apocalypse, a story entitled "Ecocatastrophe." The population question also powered the impressive pessimism of the award-winning science fiction writer John Brunner. After his 1968 *Stand on Zanzibar*, a novel of the failure of predictive science in the face of overpopulation, he produced the 1972 book *The Sheep Look Up*, the grimmest vision of ecocatastrophe to date. Brunner combines Carson's technique of composite portraiture with the traditional extrapolation of futurist fiction to show what will happen if everything that can go wrong does go wrong at the same time in an overpopulated and unevenly developed technological world.

The Difficulty of Keeping Millennialist Hope Alive in Activist Literature

Is it naive to ask why such a despairing vision of life has prevailed in ecological literature, both fiction and nonfiction? Of course, beyond the very real possibility of mass suicide by nuclear war or environmental poisoning, there is the discontent of civilization that Freud taught us lives within the hearts of us all in the developed world. Apocalyptic narrative expresses our perverse wish to destroy that which gives us comfort, the industrial society that pollutes at the same time that it improves our lot. The expressed discontent of the apocalyptic vision may even support the very social structures it apparently condemns, providing the occasion for what the social critics of the 1960s, overlaying Sigmund Freud with Karl Marx, taught us to call "repressive desublimation," the use of literature as a safety valve to release emotional steam that might otherwise explode into political unrest.

But the continuing recourse to apocalyptic narrative may also be an allegory for how difficult it is to sustain a fully ecological experience of life. Such a life requires a constant acceptance of experiment, failure, starting over, and partial success. As long as environmentalism seeks the absolute success promised by the myth of progress interpreted as human perfectibility, there will be chronic disappointment and discouragement—a daily dose of the end of the world. If both hopes and discouragements can be sanely limited, however, if success can be viewed as always local and limited, and if failure is similarly contained, the story of environmental progress may not be that different from the story of developmental progress.

This discovery powers the recent work of the journalist Bill McKibben, who had celebrated the twentieth anniversary of Earth Day in 1990 with his forlorn book *The End of Nature,* which argues that every part of the world, its land, and its atmosphere now exhibits the presence, often the destructive presence, of humankind. His 1995 book *Hope, Human and Wild* presents a chastened outlook, one that admits the need for hope and the need to search for stories of success. Unlike Gregg Easterbrook, whose 1995 *A Moment on the Earth* promotes an unchastened optimism as a response to new evidence of environmental recovery, McKibben accepts a tentative and cautious approach to the possibility of success:

> Wilderness—in its truest sense, of places totally separated from human influence—is extinguished. . . . But I'm done mourning. Innocence gone, we need to work wisely to build societies that allow natural recovery, let the rest of creation begin, however tentatively, to flourish once more. (15)

The book is a series of stories about the rebirth of natural creativity in surprising places—the overpopulated northeastern United States and the poor cities of Brazil and India—and about the need to search for stories among diverse populations. In a world dominated by Western media images, McKibben argues, "socio-diversity" is as much at risk as "biodiversity," so much so that the elements of a new narrative are difficult to find: "Once there were different ideas about time, places, limits, and responsibility to neighbors. And these differences were vast—the gulf between a cyclical and a linear concept of time, for instance, is far more profound than the gulf between communism and capitalism" (53–54). Collecting images and stories—building a fund of diverse rhetorical elements and thereby enriching the environmental imagination—thus becomes an activist practice.

In his introduction, McKibben speaks of what the classical rhetoricians called *kairos,* the sense of knowing when and how to speak, the grasp of timeliness and aptness: "Like a comet orbiting the earth, widespread concern for the environment swings into view only periodically. The time of Rachel Carson, culminating in the first Earth Day, for instance, or the furor over the greenhouse effect that peaked on Earth Day's twentieth anniversary" (2). These are the times when large-scale political action are possible and, McKibben suggests, when the most powerful rhetorical appeals, such as the apocalyptic narrative, are effective. In the lag times between the peaks of interest and action "come quieter moments for laying plans, for collecting the ideas and solutions that will form the basis of the next environmental debate" (2).

It is to the great credit of Rachel Carson that she encompassed in the space of a single work the narrative foundations of both high-energy activism and careful planning. *Silent Spring* has been read primarily as an activist document in the apocalyptic tradition, but it rewards rereading in the quieter, calmer moments that McKibben mentions. The same is true of a novel like Le Guin's *Left Hand of Darkness,* which seems still relevant to contemporary ecological concerns, while novels like her own *Word For World Is Forest* or Herbert's *Green Brain* seem at best quaint and dated by comparison, period pieces from a time when apocalyptic fire burned green.

If Carson had given over totally to apocalyptic rhetoric, she would have forsaken her scientific outlook on life. There may be no completely happy endings in the master narrative of scientific research, but neither is there a scene of total destruction. There are, as in tragedy, the signs of rebirth and continuance. The work continues, the search goes on. The same is true of the tentative treaty with which the action ends in Le Guin's *Left Hand of Darkness* and the uncertain victory of the hero Paul and his Fremen supporters in Herbert's *Dune*. As John Dewey suggested in his great treatise on how modern science alters our philosophical understanding, the quest for knowledge is always open-ended and uncertain. "To discover and illuminate truth" is the aim of both science and literature, as Rachel Carson said (qtd. in Brooks 128), but the truth of narrative, like the truth of science, is never absolute; it always points toward the next experiment, the next story. Although the expression of pure science in the conference papers and journal articles of specialized disciplines like physics, biology, and even ecology is usually understood as antimythological and nonnarrative (see Lyotard 25–27), the conversion of science to purposes of human action may depend upon translation into a generally comprehensible narrative framework and absorption into the mythology of the nonexpert citizen. Toward this end, which concerns both science fiction and literary nonfiction on scientific themes, Rachel Carson left a distinctive mark, one that still serves today as a memorable guidepost.

Appendix
Narrative Elements in *Silent Spring*

Chapter and Title	Contribution to Overall Narrative	Sample Anecdotes or Episodes, Heroes, and Villains
1. A Fable for Tomorrow	Overture: the people bring death upon themselves.	None.
2. The Obligation to Endure	Humankind has claimed the power to alter nature within one generation—just as surely with pesticides as with nuclear war: hubris unchained. Agricultural productivity overrides the warnings of ecologists.	Ecologists sound the warning (e.g., Charles Elton, Paul Shepard).
3. Elixirs of Death	The "sudden rise and prodigious growth" of the postwar pesticide industry in league with agri-business threatens all organic life: hubris clobbered by nemesis.	Heroes sound the warning (e.g., cancer researcher W. C. Hueper), but the poisons, from arsenic to chlorinated hydrocarbons, win the day. Müller wins the Nobel Prize for discovering DDT's insecticidal properties. DDT, chlordane, heptachlor, etc. take on a life of their own; alien substances that threaten life, cross interfaces, invade biology. Innocent creatures, from cattle to pets and human babies, die. Carson establishes mythological status with comparisons to Medea's robe, the brothers Grimm, Charles Addams' cartoons.
4. Surface Waters and Underground Seas	Pesticides pollute the water, creating another avenue for death: hubris clobbered again.	A few heroic scientists warn us (e.g., Eliassen, Hueper) with little effect. The demon DDT kills fish. Chemicals from a manufacturing plant in Colorado "travel" to a farm

		ing district via "a dark, underground sea . . . to poison wells, sicken humans and livestock, and damage crops." Runoff kills waterfowl in northern California.
5. Realms of the Soil	The biologically rich but "delicately balanced" soil is threatened by high concentrations of pesticides, which it passes on to vegetables: hubris clobbered again.	A few heroic studies warn, but . . . tobacco ruined by arsenic, California and South Carolina sweet potatoes by BHC, northwestern hops by heptaclor.
6. Earth's Green Mantle	Earth's vegetative cover, which contain many vital links in the ecological "web of life," is carelessly overkilled by herbicide use, but the alternative of biological control appears on the horizon.	The chain of destruction claims the sage grouse and the antelope in the Rocky Mountain region. In Connecticut, road building leads to the stripping bare of "nature's own landscaping." The heroic warnings sound (e.g., William O. Douglas). Ecologists (e.g., Frank Egler) recommend alternative spraying techniques that are more effective and natural and use of biological controls (e.g., introduction of helpful insects).
7. Needless Havoc	Heroes (professional wildlife biologists on the scene of the destruction) face off with villains (industry-influenced scientists and agencies who deny the scale of destruction)—the cover-up and the exposé: hubris challenged— nemesis with a human face.	"Drastic" spraying for Japanese beetle is defended by Michigan state agencies but declared unnecessary by naturalists (e.g., W. P. Nickell). Results are reported by birdwatchers and scientists alike. The episode is repeated in Illinois and elsewhere.
8. And No Birds Sing	The people awaken. The birds are disappearing, a harbinger-symbol perhaps of our own fate: nemesis made clear.	Ordinary people (e.g., "housewives" and other heroic "observers") join with professional ornithologists in sounding their dismay and their warning: episodes from all around the country (the book's central, anecdotally rich chapter). An angry conclusion convicts a generalized villain—"the authoritarian temporarily entrusted with power" who has ordered massive spraying "during a moment of [public] inattention."

9. Rivers of Death	Still the destruction continues, as fisheries are threatened.	Canadian salmon fisheries are nearly destroyed by DDT. Situation repeats in Maine, Montana, Louisiana, Florida, and elsewhere. Wildlife biologists (e.g., H. R. Mills) warn, fishermen lament, but . . .
10. Indiscriminately from the Skies	The destruction continues as the threat spreads, bolstered by an "end-justifies-the-means" philosophy of agricultural productivity, while citizens begin to resist and "more sane and conservative methods" of insect control are introduced.	The U. S. Department of Agriculture campaigns against the gypsy moth in the Northeast (an anecdotally rich chapter). Citizens resist in New York. Fire ant campaign in the South. "Sales bonanza" for chemical companies. Government collaborates with industry "propaganda" and ignores "effective and inexpensive methods of local control."
11. Beyond the Dreams of the Borgias	The extent of pesticide power increases as the "average citizen" is lulled into acceptance by "the soft sell and the hidden persuasion": hubris thrives through propaganda. The solution? Tighten government controls on pesticide use, pursue nonchemical methods.	DDT is everywhere; while the native diet of peoples like Eskimos on the Arctic shores of Alaska remain chemical-free, those who have contact with "civilization" return home tainted with residue in their fat cells. Cases of chemical industry violation of limits and regulations proliferate.
12. The Human Price	Nemesis strikes home. Human beings as part of nature become threatened—but the poisoning happens quietly, working slowly on the complex "ecology within our bodies."	Heroes strike the warning (e.g., David Price, René Dubos). Stories of farm workers and others affected by pesticides, diseases of the liver, nervous system, organic phosphate poisoning, Jamaica ginger poisoning— all complicated by the slow and often hidden manifestation of symptoms.
13. Through a Narrow Window	Nemesis strikes home: pesticides threaten oxidation systems within the human body. Genetic mutations are a possible but not proven effect of exposure.	Geneticists show how chemicals affect human genetic structure (e.g., H. J. Muller, Macfarlane, Alexander), though some scientists demur.
14. One in Every Four	Nemesis strikes home: the cancer threat. Our only hope is to identify and eliminate cancer-causing agents we've introduced into the	A historically and anecdotally rich chapter. Cancer researchers begin heroically to demonstrate dangers in man-made carcinogens (e.g., Hueper,

	environment, just as we have eliminated certain types of germs that threatened nineteenth-century life.	Hargraves). Victims appear—innocent workers, women, children.
15. Nature Fights Back	Nemesis strikes home: "the final irony." Problems with "hordes" of pests when the "balance of nature" is upset. The solution: work with nature, develop biological controls and selective spraying.	Stories abound: coyotes are destroyed and field mice proliferate; deer overpopulate when predators are destroyed to protect cattle; certain insects are more successful in resisting insecticides than their predators (other insects or birds). We "trade one enemy for another," time and again.
16. The Rumblings of an Avalanche	Nemesis strikes home: acquired immunities and resistance among insect populations. The solution: light and highly selective spraying only with demonstrated necessity and as a last resort.	The blue tick in South Africa, the malaria-bearing mosquito, and many others. Entomologists outside of industry (e.g., A. W. Brown) sound the warning.
17. The Other Road	The cautiously hopeful conclusion: alternatives to chemical control promise better times ahead; a creative science ought to prosper in place of the "Neanderthal Age of biology" with its philosophy of controlling nature.	Heroes abound (e.g., Knipling, Muller) in stories of insect sterilizations (e.g., the screwworm); "weapons forged" from the "insect's own life processes" used in insect traps (e.g., the gypsy moth), predators introduced, etc.

Notes

1. Carson was not even a science fiction reader and probably shared her critics' low esteem for the genre. In 1959, while in the midst of researching and writing *Silent Spring*, when asked by the editor of the 1956 reference book *Good Reading* for some science fiction titles to fill out her entry on biology in the new edition, she replied by saying that she knew next to nothing about science fiction and that, moreover, the few books she had seen were not worthy of mention. We are grateful to Carson biographer Linda Lear for telling us about this unpublished letter.

2. That Carson was the most important writer in the development of a science-based ecoactivist literature is a point we have developed in our earlier work, both in our book *Ecospeak* and in two articles, "The Discourse of 'Envi-

ronmentalist Hysteria'" and "Millennial Ecology." In this essay, we build upon our previous work. In *Ecospeak* and "Millennial Ecology," we lay the groundwork for understanding the two master narratives explored in the current essay—the apocalyptic and the millennialist—while in "Discourse," we follow a method similar to the one we try here, using critical comments about environmentalist writing, especially that of Carson, to explore parallels in seemingly distant discourses. There we are interested in the relationship of psychoanalytic to ecological narratives. Here we are interested in the relationship of science-based ecoactivist writing to science fiction.

3. The pages of Le Guin's introduction are unnumbered in all editions.

4. The findings of the Club of Rome are compiled in Meadows et al.

5. On the concept of a master narrative, or metanarrative, see Lyotard. Our use of narrative analysis as an approach to rhetorical criticism is inspired by Fisher.

6. The idea that DDT and other chemicals become "characters" was suggested to us by Randy Harris. His excellent essay in this volume explains another, dialogistic dimension of character in the treatment of how "voices" are introduced into the text.

7. We are indebted to Prof. Elisa Key Sparks of Clemson University for her suggestions about science fiction texts relevant to our study. Prof. Sparks has been studying environmental themes in science fiction for a number of years now, and her advice was very generous and helpful.

8. Herbert ultimately forsook the environmentalism of his early, best work. In the sequels to *Dune*, the heroic side of heroic millennialism, a concern with the modern messianic tradition that brought us the likes of Adolf Hitler and Joseph Stalin, becomes his obsession. In a 1970 essay, Herbert says that he had hoped that *Dune* "would be an environmental awareness handbook" (*Maker* 249), but by 1983, he balked at being associated with the environmental movement:

> Occasionally, I'm identified as an ecologist. People don't realize that I'm not a hot-gospel ecologist. . . . Ecology has become, rather deservedly, a dirty word. Because it has been picked up by a lot of demagogues and a lot of people who are not looking at all of the necessities of their time. (*Maker* 158)

He explains in a 1980 piece that his journalism on ecology prior to writing *Dune,* journalism that led him to dedicate the book to the dryland ecologists of his own time, also led him to think that "ecology might be the next banner for the demagogues and would-be heroes, for the power seekers and others ready to find an adrenaline high in the launching of a new crusade" (*Maker* 99).

Works Cited

Aldiss, Brian W., with David Wingrove. *Trillion Year Spree: The History of Science Fiction*. New York: Avon, 1986.

Bainbridge, William Sims. *Dimensions of Science Fiction*. Cambridge: Harvard UP, 1986.

Bakhtin, M. M. *The Dialogic Imagination*. Trans. Caryl Emerson and Michael Holquist. Ed. Michael Holquist. Austin: U of Texas P, 1981.

Bosso, Christopher J. *Pesticides and Politics: The Life Cycle of a Public Issue*. Pittsburgh: U of Pittsburgh P, 1987.

Brooks, Paul. *The House of Life: Rachel Carson at Work*. Boston: Houghton, 1972.

Buell, Lawrence. *The Environmental Imagination: Thoreau, Nature Writing, and the Formation of American Culture*. Cambridge: Harvard UP, 1995.

Callenbach, Ernest. *Ecotopia*. New York: Bantam, 1975.

Carson, Rachel. *Silent Spring*. Boston: Houghton, 1962.

Clarke, Arthur C. *Childhood's End*. New York: Harcourt, 1953.

Dewey, John. *The Quest for Certainty*. New York: Putnam's, 1929.

Easterbrook, Gregg. *A Moment on the Earth: The Coming Age of Environmental Optimism*. New York: Viking, 1995.

Fisher, Walter R. *Human Communication as Narration: Toward a Philosophy of Reason, Value, and Action*. Columbia: U of South Carolina P, 1987.

Freeman, Martha, ed. *Always, Rachel: The Letters of Rachel Carson and Dorothy Freeman, 1952–1964*. Boston: Beacon, 1995.

Graham, Frank, Jr. *Since Silent Spring*. Boston: Houghton, 1970.

Herbert, Frank. *Dune*. New York: Ace, 1965.

———. *The Green Brain*. New York: Ace, 1966.

———. *The Maker of Dune: Insights of a Master of Science Fiction*. Ed. Tim O'Reilly. New York: Berkley, 1987.

James, Edward. *Science Fiction in the 20th Century*. New York: Oxford UP, 1994.

Killingsworth, M. Jimmie, and Jacqueline S. Palmer. "The Discourse of 'Environmentalist Hysteria.'" *Quarterly Journal of Speech* 81 (1995): 1–19.

———. *Ecospeak: Rhetoric and Environmental Politics in America*. Carbondale: Southern Illinois UP, 1992.

———. "Millennial Ecology: The Apocalyptic Narrative from *Silent Spring* to *Global Warming.*" *Green Culture: Environmental Rhetoric in Contemporary America.* Ed. Carl G. Herndl and Stuart C. Brown. Madison: U of Wisconsin P, 1996. 21–45.

Knight, Damon. "Natural State." *Three Novels.* 1952. New York: Doubleday, 1967. 82–150.

Lear, Linda. "Rachel Carson's *Silent Spring.*" *Environmental History Review* 17.2 (1993): 23–48.

Le Guin, Ursula K. *Always Coming Home.* New York: Harper, 1985.

———. *The Left Hand of Darkness.* 1969. New York: Ace, 1976.

———. *The Word for World Is Forest.* New York: Berkley, 1972.

Lyotard, Jean-François. *The Postmodern Condition.* Trans. Geoff Bennington and Brian Massumi. Foreword by Frederic Jameson. Minneapolis: U of Minnesota P, 1984.

McKibben, Bill. *The End of Nature.* New York: Doubleday, 1989.

———. *Hope, Human and Wild: True Stories of Living Lightly on the Earth.* Boston: Little, 1995.

Meadows, Donella H., Dennis L. Meadows, Jørgen Randers, and William W. Behrens III. *The Limits to Growth.* New York: Universe, 1972.

Robinson, Douglas. *American Apocalypses: The Image of the End of the World in American Literature.* Baltimore: Johns Hopkins UP, 1985.

Scholes, Robert. *Structural Fabulation: An Essay on Fiction of the Future.* Notre Dame: Notre Dame UP, 1975.

Afterword:
Searching for Rachel Carson

LINDA LEAR

When Rachel Carson died in April 1964 at the age of fifty-six from complications of rapidly metastasizing breast cancer and heart disease, she had already become an icon, bereft of the passion and wonder of life that had been part of her personal triumph. Much of the iconic image and lore about her was forged in the vituperative controversy that erupted when *Silent Spring* was first serialized in the *New Yorker* in June 1962 and continued after the book appeared in September. The image of Carson as saint deepened during the first decade after her death when the newly emerging environmental movement appropriated her goals and made her its patron. The first generation of historians, journalists, and biographers further promoted Carson's image as a remote but heroic reformer.

Central to the Carson image is the notion that Rachel Carson was a private person who had no life other than her work, so there was no need to probe into her person further. The outlines of this icon are that she was a simple spinster who lived with her mother for over fifty years, loved nature, and fancied cats. Financially and emotionally burdened by her family's care, she spent fifteen years in a dead-end job as a government science editor, eking out time for her own writing.

Although shy and retiring, Carson had a special relationship with the natural world and a gift for writing about it. A trilogy of books on the sea made her famous and allowed her the luxury of a summer home in Maine. In the early 1960s, concern about the excesses of applied chemical pesticides prompted her to write *Silent Spring,* an uncharacteristically polemical book of warning, perhaps motivated by her own illness. She died, according to this fanciful version, satisfied with what she had begun. In sum, Carson's life was circumscribed and often tragic, but her legacy endured.

Rachel Carson consciously contributed to this impersonal image. She was genuinely reserved and fiercely protective of her personal privacy and that of her family. She had firsthand experience with the public's curiosity about the private lives of famous people. The popularity of *The Sea Around Us,* published in 1951, gave Carson an international reputation. When the spotlight came again in 1962, she was determined to keep her family secrets private and to protect her grandnephew Roger Christie, whom she adopted after the death of his mother in 1957. Anticipating the public interest that *Silent Spring* would provoke, Carson took pains to remove herself from the controversy. The image she had created served her purposes.

Her advancing cancer, manifested by what she called a "catalogue of illnesses" after 1960, compounded her personal need for privacy. Carson was determined to keep her declining health a secret from the press not only because she needed to be able to advocate her views objectively but also because she knew her critics would use it to discredit her evidence. Her illness prevented her from accepting many of the awards and honors that came to her and from public appearances that she longed to make. *Silent Spring*

was a political powder keg, and Carson was well aware of its political, economic, and legal ramifications. Desperately ill through 1963,
Carson—together with her agent—carefully orchestrated her public appearances and limited her exposure to the media, adding to
her remoteness (Lear, *Rachel Carson* 429–41).

The controversy and acclaim that greeted *Silent Spring* put Carson in a category usually reserved for the most vaunted literary
figures. William O. Douglas, associate justice of the United States
Supreme Court, described her as a "prophet in her own time" and
declared the book "the most important chronicle of this century for
the human race." Just after Carson died, John Kieran, the naturalist
and author, told a reporter that he had always thought of her as a
mystical being, calling her a "nun of nature" (Douglas; "Rachel
Carson Dies").

Rachel Carson dedicated *Silent Spring* to Albert Schweitzer,
whose philosophy she revered. Carson always credited her mother's
reverence for life as her introduction to Schweitzer's great precept
and the basis of her attitude toward nature. Her awareness of the
stream of life, as well as her belief in the principle of a balance in
nature, reinforced the view that Carson had, as Schweitzer before
her, a mystical connection with the natural world that set her apart.
Critics seized on this aspect of her work, fearing "her emotional
and inaccurate outburst . . . would do no good for the things that
she loves" ("Pesticides").

While Carson was privately annoyed by such assessments, her
silence in the face of outrageous personal attacks by government
and industry and her composure in public debate only reinforced
her posthumous stereotype as "Saint Rachel." Former Secretary of
Interior Stewart Udall, who knew Carson and became her cham-
pion during the years of controversy over *Silent Spring,* added to the
icon when he called her the "fountainhead of the modern environmental movement" in a speech shortly after Earth Day in 1970
(Udall, Personal interview, Speech). The Environmental Protection

Agency put her picture on the cover of its first monthly magazine. In 1980, her image graced the new seventeen-cent U.S. postage stamp. She was awarded the Presidential Medal of Freedom posthumously in 1980. The citation read, in part, "Always concerned, always eloquent, she created a tide of environmental consciousness that has not ebbed." She had become a mystical, bloodless revolutionary, whose writing was widely quoted, whose spirit was invoked at every political rally of environmental activists, but whose personal and private life was empty and, by implication, uninteresting.

The first generation of historians who reviewed her contributions confirmed this assessment. Finding nothing in her personal life worthy of comment, they pigeonholed her as a romantic in the canon of environmental literature, conveniently ignored her penetrating scientific and cultural critique, and trivialized her literary achievements. Donald Fleming, Stephen Fox, and even Roderick Frazier Nash, dean of American environmental history, treat her like a road sign along the one-way highway that connects historical labels. Fleming, writing in 1972, took a simplified, linear look at Carson's writing, dismissing her as a mere summarizer of science. He suggests that she deliberately subordinated the broader theme of harm to the basic ecology in order to emphasize the menace to human health. He calls *Silent Spring* an aberrant book in the body of Carson's writing that went "against the grain of her temperament." Her distinctive role, he claims, was to renew the romantic notion of self-obliteration before nature; thereby, he absolves later interpreters from having to deal with a complex writer who left a challenging literary and political legacy (11–12).

Almost a decade later, Stephen Fox, writing in his *American Conservation Movement*, notes that, after early writing "squarely in the literary naturalist mode," with *Silent Spring*, Carson initiated the "burgeoning of man-centered conservation." According to Fox, Carson herself "illustrated the shift from traditional conservation to a focus on human welfare," an impersonal accomplishment by a woman suspicious of the direction of modern technology (292–93).

Roderick Nash, in *The Rights of Nature* (1989), notes with evident surprise that in a male-dominated field, the two books that proved the most effective in extending American ethics were the work of women, Harriet Beecher Stowe's *Uncle Tom's Cabin* and Rachel Carson's *Silent Spring*. Carson's work, Nash writes, is "a landmark in the development of an ecological perspective. It did much to accelerate the new environmentalism and generated the most widespread public consideration of environmental ethics to that date." "Her forte," he said, "was not original research, but old-fashioned natural history, colored in every respect with something many modern ecologists had forgotten: love of nature" (78).

Dismissing Carson's intellectual contributions, other male biographers, including Stephen Fox, obsessively focused on Carson's marital status. Noting, incorrectly, that Carson lived with her parents all her life and that the family circle widened to include two nieces and a grandnephew, Fox suggests that Carson never married because her family responsibilities exacted the price of isolation from the outside world. In dress and manner, says Fox (who never knew her), Carson was "neat, quiet, and old fashioned." "Carson would have liked to marry," he theorizes, but was apparently "inhibited by the familial cocoon." Isolated and overprotected, Carson did nothing to encourage a man's romantic interest in her, Fox surmises. Since it was apparently not possible for him to imagine that a woman might choose *not* to marry, he searched for larger personal reasons and came up with a hardship and surplus theory. In the 1930s, many men could not afford to marry, while in the 1940s, many men were at war, leaving a surplus of women in Washington. According to Fox, poor Rachel was an unfortunate victim of circumstances (Fox 294).

Fox's view drew on the work of Philip Sterling, who also never knew Carson but who interviewed some of her family and childhood friends. In his biography for young adults, *Sea and Earth: The Life of Rachel Carson,* published in 1970, Sterling managed to project on Carson the full set of sexist stereotypes about single, professional women. Commenting on Carson's studiousness as an undergraduate, Sterling writes, "Why worry about a girl who was so different? If she wanted to spend her weekends in the library studying and in her room. . . . Oh well, that was her business" (43). He portrayed Carson's life without a husband as mostly sad rather than as a life happily filled with relationships that she took pleasure in choosing. According to Sterling, Carson was dogged by the shadow of loneliness and unfulfillment. He describes her as a "quiet little government employee" who succeeded through pluckiness, good grooming, and good manners, only to remain forever the invisible little girl. Even when describing Carson's testimony before Congress in June 1963, Sterling infantilizes her: "Sitting at the long, witness table, she looked small, middle-aged, harmless. Her appearance gave no hint that she might indeed be capable of starting something big enough to interest the Senate of the United States" (1). Dismissing Carson's ten-year relationship with Dorothy Freeman, a neighbor in Maine, Sterling calls Carson a "lifetime loner." For him, as for most of the other commentators, female friendships are insignificant. Carson's many important friendships with women—with her mentor, biology professor Mary Scott Skinker; with her literary agent, Marie Rodell; and later with a coterie of devoted female friends—are not interesting. They are relationships by default, not by choice.

Even Carson's editor at Houghton Mifflin, Paul Brooks, inadvertently contributed to the icon. Brooks's 1972 account, *The House of Life: Rachel Carson at Work,* was conceived and authorized by Marie Rodell as a reader, with lengthy selections of Carson's writing introduced by biographical notes to put them in context. Rodell and Brooks agreed that Carson would not wish a biography and hoped that a book on Carson and her work would satisfy public interest and put an end to such drivel as that proffered by Sterling. Although Brooks's book sold adequately, it was never the success that Brooks and Rodell had hoped. It too portrayed Carson as a disciplined and gifted writer who suffered from too many responsibilities and too little joy.

Brooks, an activist and winner of the John Burroughs Award, also added to Carson's ascetic image when he described her attitude toward the natural world as "always that of a deeply religious person." Despite the fact that he respected Carson's literary and scientific talent, admired her courage and discipline, and felt compassion for her personal burdens, Brooks echoed the prevailing male prejudice that an unmarried woman lacked fulfillment. His explanation for Carson's status was, however, more plausible than those offered by other commentators: her mother kept her from marrying. Acknowledging Maria Carson as the single most important influence in her daughter's life, Brooks writes that Carson's "family responsibilities, whatever their rewards and satisfactions, kept Rachel from enjoying what Thoreau called a broad margin to her life. And it is probably an understatement to say that Maria Carson never urged Rachel to marry" (242).

As Patricia Hynes has pointed out in her maverick study, *The Recurring Silent Spring,* when Thoreau wrote that he required "broad margins to his leisure," it was by way of expressing his need for privacy and time for reflection. When Brooks used this phrase, he was not referring to Carson's requirements as a writer but to the barrenness of her life as a single woman (Hynes 64). Although Brooks knew Carson well, he judged her life, and especially her emotional happiness, by his own gender-based values and his paternalistic desire to protect her. He could not fully appreciate Carson as an independent woman of richness and complexity who chose not to marry in order that her creative life might be more fulfilling. Until recently, only Hynes seemed troubled that Carson's interpreters accepted the view that an unmarried, childless woman had no life apart from her work (Hynes 60–62). It seems to me, however, that the rare unity of Carson's life and work attests to a striking integrity of spirit and mind.

Carson herself gave the best clue to any future biographer when she shared one of her own strategies as a writer: "[B]e still and listen to what the subject has to tell you." When we do that, we look for how she dealt with the unexpected twists in her life and with the choices those events elicited. One of the things we discover is that independence was central to Rachel Carson's creativity and emotional well-being. From her adolescence onward, she was aware that her life had a larger purpose. Through her role of interpreter of science for the public, she sought to leave a broader witness for nature. In carrying out her vision, Carson was sustained by the friendship and counsel of both men and women, but we now know that from 1953 until her death, she derived particular fulfillment from the love and devotion offered by Dorothy Freeman.

Writing Carson's biography presented some unique challenges. Once knowing that the icon was deficient, the critical problem became gaining access to the material from which to discover the other life Carson lived. In her will, Carson named her literary agent, Marie Rodell, as her literary executor and gave her, and Rodell's assigned heir, sole authority for future use of all her words, published and unpublished. After working with Carson for over fifteen years, Rodell knew most of her secrets, as well as her desire to be judged by her literary output. Rodell deposited Carson's thoroughly censored papers at the Yale University Collection of American Literature in the Beinecke Rare Book and Manuscript Library in 1966, where they remained an uncataloged jumble until 1990. Only Paul Brooks and Audubon writer Frank Graham, Jr.—author of *Since Silent Spring* and also one of Rodell's clients—were given permission to see any of the private papers and letters Rodell withheld, and only Brooks had access to a portion of Carson's correspondence with Dorothy Freeman. Even Rodell had an incomplete knowledge of Carson's intimate relationship with Dorothy Freeman, and she seriously underestimated the extent of the correspondence between the two. Rodell's successor, Frances Collin, never knew Carson, but her loyalty to Rodell's policies was ironclad. There would be no biography, Brooks's reader was the final and official word.

This was the status of things twenty-two years after the publication of *The House of Life,* when I discovered my own intellectual debts and personal connections to the private Rachel Carson and embarked upon an independent biography. No new life of Carson could be written unless the estate gave permission to quote unpublished material and unless one could gain access to the Carson–Freeman correspondence, which was physically in the hands of the Freeman family but controlled by the Carson estate. For a variety of reasons, both entities were initially reluctant to support my efforts. So I began my search for Rachel Carson with the only uncensored sources available, Carson's friends.

In the course of nearly a decade of work, I interviewed more than one hundred twenty individuals who knew Carson well. From the beginning, Paul Brooks and Carson's colleagues Jeanne Davis, Ruth Scott, and Shirley Briggs supported my work and pressed the estate on my behalf. Among these, Paul Brooks became my chief advocate. After the estate gave me unrestricted permission, the Freeman family agreed to permit access to all of the correspondence and to publish most of it in 1995 as *Always, Rachel: The Letters of Rachel Carson and Dorothy Freeman, 1952–1964,* edited by Dorothy's granddaughter Martha Freeman.

While solving some problems, publication of the Carson–Freeman letters has also raised others. This correspondence begs, but does not resolve, questions about the nature of their relationship, and I was left to deal with sexual ambiguity in an age that lusts for categorization. The letters do, however, establish that Dorothy

Freeman was the love of Rachel Carson's life and that their relationship gave her the emotional support she needed to carry out her witness. I am also convinced that the loving relationship between the two women was more powerful than sex and that the latter is irrelevant to discovering who Rachel Carson was and what she was trying to accomplish.

Rachel Carson: Witness for Nature, exposes many of the family secrets that Rachel worked hard to hide. Insofar as they are germane to revealing who Carson was and how they impacted her work, I have told them. Carson was burdened by a troubled and difficult family, and her life and literary output were defined both by them and because of them. But times and tastes have also changed, and what was humiliating fifty years ago hardly raises an eyebrow today.

I can replace the icon with a more truly heroic, far richer, and more passionate woman than the world has thus far embraced. Rachel Carson was imbued as a young woman with an intense love of the natural world, a delight in its processes, and an unusually keen conviction that she was meant to bear witness to that wonder; she was uncertain, however, about the form that witness would take. Aware of her literary gifts at an early age and disturbed by the damage that polluting industries had wreaked upon the once-sparkling landscape of her childhood home, she found her calling as an interpreter of the science of the earth and its life. Science showed her the beauty of life and convinced her that if she could widen the circle of people engaged in wondering at it, life would be safe from the evermore-destructive forms of human arrogance.

Certain that the traditional role for women would interfere with her vision, Carson vigorously and courageously defined herself as a professional scientist and science writer who wrote for the public. Her family responsibilities and her interests decreed that for a time she had to make another career as a government scientist and editor. Her experiences there did not circumscribe her outlook; rather they offered her unprecedented professional access and opportunity. Although her personal and social activities were limited by lack of financial resources and excessive family demands, her relationships were wide and rewarding. She was most fulfilled when she was interpreting the ecology of a refuge or describing the environment of a marine organism (Lear, *Lost Woods*).

Fundamentally shaken by the use of the atomic bomb at the end of World War II, and sensitized by the existential outlook of the postwar period, Carson feared that humankind had lost its moorings, particularly with regard to its place as part of the natural world, as well as in the infinite stream of time. Passionately committed to righting this distorted view and to making the truths of science available to all, she wrote with an increasing sense of commitment after the success of *The Sea Around Us* gave her both public forum and financial freedom.

In speech after speech during the 1950s, Carson urged taking the longer view of geologic time as a way to moderate the effects of modern technology that made the living world increasingly remote to everyday life. She believed that the possibility of nuclear war, the values of a newly affluent society, and the arrogance of science and technology threatened not only the environment but also the health and welfare of those dependent upon it. As she moved in her own literary work from the study of the primal forces of the ocean to the environmental damage caused by human arrogance, Carson's scientific perspective and ethical awareness broadened as well (see Carson).

Although she had been interested in naturally occurring poisons in the environment since 1938, her editorial experience after 1945 uniquely positioned her to grasp the threat that chlorinated hydrocarbon pesticides, such as DDT, posed to humans and wildlife. She had several important books under contract in 1958, but she accepted the challenge posed by an unquestioning society and a generally smug scientific establishment to inform the public of the dangers they faced in the broadcast use of these poisons.

During the nearly five arduous years that she worked on *Silent Spring,* Carson's special relationship with Dorothy Freeman provided the emotional centering and love she needed to succeed. But Carson's witness was more important than any relationship. When Carson's cancer returned at the end of 1960, she realized on some level that *Silent Spring* might be her last book. There is no evidence that her disease influenced the conclusions that she drew in 1962, but it did propel her to a different level of witness. The book she would have published in 1960 was far different from the one that appeared in 1962. Her "catalogue of illnesses" encapsulated time and forced her to simplify her scientific explanations of pesticide chemistry and the molecular changes pesticides induced in the human cell. That same enforced period of reflection allowed her to find a way to frame her evidence so that ordinary people would listen to her message. Since she might not have another opportunity, it propelled her into uncharacteristic boldness and to televised debate with industry and government opponents in May 1963. When asked to give congressional testimony a month later, Carson seized the only moment she might have to move the subject to another level and leave a different sort of legacy.

One of Carson's greatest achievements as a public figure was in sensing how much she could reveal of the dangers of pesticides without alarming the public unduly and in calculating just how much the public could absorb. There was much she knew that she chose not to say. Yet she left a clear trail of evidence for scientists and reformers who followed her, whose times, she hoped, would permit freer disclosure. The power of her vision is impressive, and it is tragic that cancer denied her the promise of what she had learned just when she had reached the height of her analytical powers and had an enviable forum for speaking out. Rachel Carson died frustrated that she had been given so little time and had so much more to say.

Carson would not have been surprised by the words of a *Boston Globe* editorial writer who summarized her legacy after Earth Day in 1970, writing, "A few thousand words from her and the world took a new direction" ("Rachel Carson Explosion"). She had hoped to do just that, but she was realistic enough to know that the changes she intended would be long in coming and difficult to maintain. There is always a reflection of truth in any icon, and the longevity of the Carson image testifies to this. As I have lived intimately with Carson these past years and struggled to discover her, I have been reminded of a line from Berthold Brecht's *Galileo*: "Unhappy is the land that needs a hero" (scene 12). We tend to see a person like Rachel Carson as different from us, capable of a life we could not attempt. But the complexity of her life makes her more like us than different.

Yet the truth contained in the Carson icon and the distance that it places her from us remind us that she, like other heroes we might name, never broke with her beliefs and remained dedicated to something outside herself. Carson never became the enemy of her soul or of her memory. Her life and her witness had integrity, and it is the compelling integrity of her witness that leads to the central challenge of Carson's life.

Brecht's line, I think, can be read in two ways: one holds that a land requiring heroes must be a sorry place; the other concludes that the emergence of heroes is a sign of trouble. I have embraced the latter position in writing Carson's life. I believe she saw postwar America as a troubled place, a land laid siege to, besieged not by greed alone but by venal imaginations that drove a potentially destructive technology and by an insidious tendency in the culture to commodify all of the living world of which human beings were but one part. Carson gave this abuse a historical context, protested its effects, warned against ignoring it, and offered a new ethic and a practical sort of hope. In doing so, she became our great citizen-scientist and citizen-writer. She was a risk taker, a reluctant but bold revolutionary. Her witness urges us to take risks and support the risk taking of others. It suggests new ways to build community and reminds us that our fate, and the fate of our planet, does not lie with a few heroes or icons but ultimately with ourselves.

Works Cited

Brecht, Berthold. *Galileo.* Trans. Charles Laughton. New York: Grove, 1966.

Brooks, Paul. *The House of Life: Rachel Carson at Work.* Boston: Houghton, 1972.

Carson, Rachel. "Remarks at the Acceptance of National Book Award for Nonfiction." Lear, *Lost Woods,* 90–92.

Douglas, William O. "Silent Spring." *Book-of-the-Month Club News* Sept. 1962, 2.

Fleming, Donald. "Roots of the New Conservation Movement." *Perspectives in American History* 6 (1972): 7–91.

Fox, Stephen. *The American Conservation Movement: John Muir and His Legacy.* Madison: U of Wisconsin P, 1981.

Freeman, Martha, ed. *Always, Rachel: The Letters of Rachel Carson and Dorothy Freeman, 1952–1964.* Boston: Beacon, 1995.

Hynes, H. Patricia. *The Recurring Silent Spring.* New York: Pergamon, 1989.

Lear, Linda, ed. *Lost Woods: The Discovered Writing of Rachel Carson.* Boston: Beacon, 1998. 90–92.

———. *Rachel Carson: Witness for Nature.* New York: Holt, 1997.

Nash, Roderick Frazier. *The Rights of Nature: A History of Environmental Ethics.* Madison: U of Wisconsin P, 1989.

"Pesticides: The Price of Progress." *Time* 28 Sept. 1962: 48.

"Rachel Carson Dies: Her Cause Lives On." Obituary. *New York Herald Tribune* 16 Apr. 1964, 30.

"Rachel Carson Explosion." *Boston Globe* 17 Mar. 1970, 14.

Sterling, Philip. *Sea and Earth: The Life of Rachel Carson.* New York: Crowell, 1970.

Udall, Stewart L. Personal interview. Falls Church, VA. 8 Mar. 1992.

———. Speech. Texas Christian U. Fort Worth, TX. 25 Apr. 1970.

Contributors

PAUL BROOKS was for many years Rachel Carson's editor and friend. He was a former editor in chief at Houghton Mifflin Co. and the author of *Roadless Area* (1964), *The House of Life: Rachel Carson at Work* (1972), *Speaking for Nature* (1980), and other books.

EDWARD P. J. CORBETT was a professor of rhetoric and composition at The Ohio State University. He was the author of *Classical Rhetoric for the Modern Student* (1965, 1971, 1990, 1999) and numerous other scholarly books and articles on rhetoric and composition. He was a founder of the Rhetoric Society of America and an editor of *College Composition and Communication*.

CAROL B. GARTNER is a professor of English at Purdue University Calumet. She is the author of *Rachel Carson* (1983), a critical appraisal of Carson's life and work, and of numerous scholarly articles and the entries on Rachel Carson in Ungar's *American Women Writers* and *Collier's Encyclopedia*.

CHERYLL GLOTFELTY is an associate professor of literature and environment at the University of Nevada, Reno. She was cofounder and past president of the Association for the Study of Literature and Environment and is coeditor of *The Ecocriticism Reader* (1996).

RANDY HARRIS is an associate professor of rhetoric at the University of Waterloo; he is the author of *The Linguistic Wars* (1993), the editor of *Landmark Essays on the Rhetoric of Science* (1997), and the author of numerous scholarly articles on rhetoric of science.

M. JIMMIE KILLINGSWORTH is a professor of English and director of graduate studies in the Department of English at Texas A&M University. He is coauthor with Jacqueline S. Palmer of *Ecospeak: Rhetoric and Environmental Politics in America* (1992) and of some of the landmark essays in rhetoric and the environment.

LINDA LEAR is a research professor of environmental history at George Washington University and a research collaborator at the Office of the Smithsonian Institution Archives. She was the primary historian for the 1993 PBS *American Experience* documentary "Rachel Carson's *Silent Spring,*" is the author of the acclaimed recent biography *Rachel Carson: Witness for Nature* (1997), and is the editor of *Lost Woods: The Discovered Writing of Rachel Carson* (1998).

RALPH H. LUTTS is a member of the faculty of Goddard College and of the adjunct faculty of the University of Virginia and Virginia Tech. He is the author of *The Nature Fakers: Wildlife, Science & Sentiment* (1990), the editor of *The Wild Animal Story* (1998), and a recipient of the Forest History Society's Ralph W. Hidy Award.

CHRISTINE ORAVEC is a professor of rhetoric at the University of Utah. She is widely published in rhetoric and women's studies and has produced some of the landmark essays in rhetoric and the environment.

JACQUELINE S. PALMER teaches technical writing at Texas A&M University, where she also serves as associate director of Writing Programs in the Department of English. She is coauthor with M. Jimmie Killingsworth of *Ecospeak: Rhetoric and Environmental Politics in America* (1992) and of some of the landmark essays in rhetoric and the environment.

MARKUS J. PETERSON is the Upland Wildlife Program leader for the Texas Parks and Wildlife Department and has authored numerous scholarly articles on wildlife ecology and environmental policy. He has coauthored with Tarla Rai Peterson several essays on environmental communication and environmental ethics.

TARLA RAI PETERSON is an associate professor in the Department of Speech Communication and a research associate in the Center for Science and Technology Policy and Ethics at Texas A&M University; she is the author of *Sharing the Earth: The Rhetoric of Sustainable Development* (1997), coeditor of *Communication and the Culture of Technology* (1990), and the author of numerous scholarly articles and chapters on rhetoric of science; like several other contributors to this volume, she has authored some of the landmark essays in rhetoric and the environment.

CRAIG WADDELL is an associate professor of rhetoric at Michigan Technological University. His work has focused on environmental rhetoric and has appeared in *Philosophy and Rhetoric; Science, Technology, & Human Values; Quarterly Journal of Speech; Social Epistemology; Technical Communication Quarterly;* and *Green Culture: Environmental Rhetoric in Contemporary America.* He is the editor of *Landmark Essays on Rhetoric and the Environment* (1998).

Index